SILICA-BASED BURIED CHANNEL WAVEGUIDES AND DEVICES

Optical and Quantum Electronics Series

Series editors

Professor G. Parry, University of Oxford, UK
Professor R. Baets, University of Ghent, Belgium

This series focuses on the technology, physics and applications of optoelectronic systems and devices. Volumes are aimed at both research and development staff and engineers involved in the application of optical technologies. Graduate textbooks are included, giving tutorial introductions to the many exciting areas of optoelectronics. Both conventional books and electronic products will be published, to provide information in the most appropriate and useful form for users.

1 **Optical Fiber Sensor Technology**
   Edited by K.T.V. Grattan and B.T. Meggitt

2 **Vision Assistant Software**
   A practical introduction to image processing and pattern classifiers
   C.R. Allen and N.C. Yung

3 **Silica-based Buried Channel Waveguides and Devices**
   François Ladouceur and John D. Love

Silica-based Buried Channel Waveguides and Devices

François Ladouceur

Optical Sciences Centre,
The Australian National University,
Canberra, Australia

and

John D. Love

Optical Sciences Centre,
The Australian National University,
Canberra, Australia

CHAPMAN & HALL

London · Glasgow · Weinheim · New York · Tokyo · Melbourne · Madras

Published by Chapman & Hall, 2–6 Boundary Row, London SE1 8HN, UK

Chapman & Hall, 2–6 Boundary Row, London SE1 8HN, UK

Blackie Academic & Professional, Wester Cleddens Road, Bishopbriggs, Glasgow G64 2NZ, UK

Chapman & Hall GmbH, Pappelallee 3, 69469 Weinheim, Germany

Chapman & Hall USA, 115 Fifth Avenue, New York, NY 10003, USA

Chapman & Hall Japan, ITP-Japan, Kyowa Building, 3F, 2-2-1 Hirakawacho, Chiyoda-ku, Tokyo 102, Japan

Chapman & Hall Australia, 102 Dodds Street, South Melbourne, Victoria 3205, Australia

Chapman & Hall India, R. Seshadri, 32 Second Main Road, CIT East, Madras 600 035, India

First edition 1996

© 1996 François Ladouceur and John D. Love

Printed in Great Britain by Hartnolls Ltd, Bodmin, Cornwall

ISBN 0 412 57930 8

A catalogue record for this book is available from the British Library

∞ Printed on acid-free text paper, manufactured in accordance with ANSI/NISO.48-1992 (Permanence of Paper).

Contents

Foreword

This book has a prehistory for which I am partly responsible, and which I would like to share with the reader. When I had the honour of being chosen as one of the examiners of Dr Ladouceur's doctoral thesis, I felt that it would be lovely if this beautiful work could be made available to a wider audience. I suggested to the author and to his thesis advisor (J. D. Love) to think about publishing the material contained in the thesis as a book. The present volume proves that this suggestion did not fall on deaf ears.

Reading the finished work, I was struck how the original thesis has grown and has evolved into a concise book that no longer resembles a doctoral thesis. However, since one of the authors is also the coauthor of an authoritative work on optical waveguide theory, a high level of sophistication may well be expected.

The field of buried channel waveguides is timely, as the technology of communications networks employing wavelength division multiplexing (WDM) techniques is just about to explode. WDM requires extensive combining and splitting of light channels which can best be done by integrated optoelectronic devices, which require the optical circuitry described in this work.

The book contains descriptions of experimental techniques and of theoretical material, with a strong emphasis on theory. The presentation is a good mix of simple derivations and physical explanations of principles wherever feasible. Where a subject would require too much background information, or where a derivation would be too lengthy for the framework of this book, the reader is referred to the literature where the required background information can be found. This strategy helps to produce a very readable, easy-to-follow text and ensures that the reader is never overwhelmed by unexplained facts, as happens in books that try to present complicated technical material without adequate explanations.

Ladouceur and Love have succeeded in making complicated concepts of electromagnetic wave propagation easily understandable. By their very nature, the building blocks of buried channel waveguides are not accessible to exact analytical solutions. For this reason, it is necessary to resort to approximate methods for their analysis. What makes the approximate analysis of buried channel waveguides possible is the concept of weakly guided waves, which is based on the fact that the refractive index of the material of the waveguide core is only slightly different from the surrounding cladding material. However, even the equations resulting from the weak-guidance approximation cannot be solved exactly. The authors have a knack for extracting useful approximate analytical results from complicated mathematics. For more complete results, they reach for their large arsenal of numerical methods for analysing more complicated structures. All numerical methods are thoroughly explained without boring the reader with too much technical detail. For the nitty gritty, needed to write a numerical analysis program, the reader is referred to the literature.

The coverage of the book is limited to the most fundamental optical circuits that can be made from buried channel waveguides. With these tools in hand the circuit designer can proceed to construct more complicated structures, such as channel-dropping filters and wavelength-dependent routers that are needed in WDM systems. The book does not cover nonlinear effects, which may possibly play a part in optical switching systems of the future. Such extensions of the basic building blocks will undoubtedly become the subject of future books.

D. Marcuse
Retired from AT&T Bell Laboratories
Visiting Research Professor at the University of Maryland

Preface

This book is designed to provide basic insight into the attributes of propagation through buried channel optical waveguides and devices, with emphasis on the weak-guidance approximation. This approximation is applicable, in particular, to waveguides and devices fabricated using silica-based processes. The book provides both a qualitative physical description and a parallel analytical and numerical quantification of the important aspects of propagation. Taken as a whole, the material presented here is self-consistent and should be sufficient to enable a competent graduate engineer or scientist to generate first designs of components and optical circuitry for specific applications.

Following the introduction, Chapter 2 provides a brief outline of fabrication techniques for planar waveguides and devices. We have not attempted to provide a complete or critical description of each of these techniques, as there is currently rapid development of existing and new processes, whereas the basic theoretical analyses presented here are likely to remain valid for a much longer period.

Chapters 3–6 cover physical, analytical and numerical aspects of modal propagation, concentrating on the straight, uniform, single-mode buried channel waveguide. We then analyse propagation losses due to splicing to fibres, material and surface roughness, substrate leakage and bending losses, described in Chapters 7–10, respectively. Chapter 11 provides various prescriptions for designing waveguide paths between points in an optical circuit which minimize bending loss. The remaining Chapters 12–14 analyse propagation through three basic planar devices, covering single-mode couplers, large-angle X-junctions and Y-junction splitters, respectively.

There is a rapidly growing body of published work in the area of silica-based planar waveguides and devices. We have not attempted to credit all major advances, instead, the references cited at the end of each chapter are those we are most familiar with, and those which have helped us understand the subject matter as presented here. Frequent reference is made to the earlier publication, *Optical Waveguide Theory* by A. W. Snyder and J. D. Love (Chapman & Hall, London, 1983), which has become the definitive waveguide theory text.

Various acronyms associated with experimental and theoretical waveguiding techniques have appeared over the years, and are used throughout this book. Each acronym and its full equivalent are listed alphabetically in the index.

F. Ladouceur

J. D. Love

Acknowledgements

We are very grateful to those organizations and many individuals who have supported and contributed to this book. The majority of the material presented here originally formed the research thesis of Dr François Ladouceur when he was a postgraduate student in the Optical Sciences Centre, for which he was awarded the Ph.D. by the Australian National University in 1992. The suggestion for writing this book originated with Dr Dieter Marcuse, then at AT&T Bell Telephone Laboratories in the USA, who has very kindly written the foreword. Much of the material presented here was generated through research contracts with Telecom Australia Research Laboratories. Their permission to use material from the associated research reports in this book is acknowledged. The authors also acknowledge the support of the Australian Photonics Cooperative Research Centre.

The optimal low-loss path designs of Chapter 11 rely on the invaluable contribution of Mr Pierre Labeye, and we are indebted to Ms Victoria Steblina for the coupler results in Chapter 12. Dr Adrian Ankiewicz performed an indispensable task in helping to check the contents of the book. We are also grateful, in particular, to Dr Antoine Durandet, Dr Simon Hewlett, Dr Peter Kemeny, Dr Tim Senden and Dr Rod Vance for their various contributions.

1
Introduction

During the last ten years, existing optical fibre and fibre-device fabrication techniques have been complemented by the development of new processes for the fabrication of planar optical waveguides and devices. These processes have evolved partly because of the ability to fabricate optical devices which are not readily feasible in fibre technology, and partly because of the very compact and highly complex optical circuitry which can be produced on a single photonic chip. They also offer the potential to integrate photonic devices with semiconductor sources and detectors to realize a compact, hybrid photonic-optoelectronic chip, complete with fibre pig-tailing. Because of their compactness and potential low cost, these types of photonic chips are attractive components for future high-capacity optical telecommunications networks now being planned.

The purpose of this book is to provide both physical insight and the necessary techniques for analysing and designing silica-based planar optical waveguides and a range of basic passive planar devices which can be incorporated into these chips. The planar technologies for fabricating these silica-based chips are based principally on either deposition-etching techniques, starting with a planar substrate, such as a silicon wafer, or ion-exchange techniques, based on a planar glass layer such as an optically flat microscope slide. Techniques based on the photorefractive effect are also being developed for the direct writing of optical circuitry using suitably doped trilayers and focussed UV beams. The terms 'planar' waveguides and 'planar' devices have evolved as a generic description, associated with the supporting sub-structure, and are now commonly used, even though the waveguides and devices normally have non-planar cross-sections.

The material presented in the following chapters is premised on silica-based planar waveguides and devices. An important property of these waveguides and devices is the slight variation in refractive index over their cross-section, typically of the order of 1% or less. This property enables a major simplification to be made in the electromagnetic analysis, whereby the full set of vector Maxwell's equations governing the six components of the electric and magnetic fields can be accurately approximated by a single scalar equation for just one component of these fields. This is the so-called weak-guidance approximation.

1.1   ADVANTAGES OF PLANAR WAVEGUIDES

Planar technologies offer several advantages over fibre technology. Firstly, it is possible to fabricate low-loss devices which are difficult to fabricate using fibres. Perhaps the best known example is the symmetric single-mode 1×2 Y-junction, which functions as a wavelength-independent 3 dB splitter. Secondly, concatenations of Y-junctions forming 1×4, 1×8 etc., splitters can be produced just as easily as single 1×2 splitters, thereby avoiding the need for splicing, which is required between successive splits in a concatenation of fibre couplers. Thirdly, very compact layouts of planar optical devices and waveguides can be designed on a single-substrate to make optimal use of a given area and so to minimize fabrication costs. This compares sharply with the large volume required to accommodate a concatenation of fibre couplers, and to minimize bend loss between successive couplers, by ensuring that the fibres are not bent too tightly.

Compared with standard single-mode optical fibres, which are extremely uniform along their length, due to the nature of the heating and drawing process from the fibre preform, transmission along single-mode planar waveguides suffers a much higher attenuation, due to roughness introduced into the waveguiding structure through the etching/deposition and ion-exchange processes. The lowest single-mode fibre attenuation is around 0.16 dB/km at a wavelength of 1550 nm, whereas even the best planar waveguides currently produced have an attenuation of the order of 0.01 dB/cm, i.e. four orders of magnitude higher. However, since the size of planar devices is typically of the order of only a few square centimetres, the total propagation loss is normally very small.

The incorporation of planar devices into fibre systems necessitates splicing between single-mode fibres and single-mode planar waveguides. Although the difference in geometry between planar and fibre guides precludes the use of standard fibre splicing techniques, such as fusion splicing, splice loss between the two can be reduced to less than 0.1 dB depending on whether passive or active techniques are employed, i.e. whether the splicing relies solely on the geometrical arrangement or whether light is transmitted through the fibre and waveguide and monitored during the splicing process.

1.2   DESIGN CONSIDERATIONS

There is a fundamental difference between the design and fabrication of fibre-based devices and planar-waveguide based devices, as can be illustrated by considering the optical coupler. Fused-taper fibre couplers have the advantage that their functional behaviour can be monitored during the fabrication process by illuminating one of the input fibres and measuring the resulting outputs, i.e. there is essentially no theoretical predesign required, and the requisite behaviour, whether a 3 dB split, wavelength multiplexing, or wavelength flattened, relies predominantly on the practical experience of the fabricator. The response of planar waveguide devices, on the other hand, cannot be monitored during fabrication, and therefore relies generally on both the theoretical design as well as the accuracy of the fabrication process. Accordingly, the simplest planar device which can be fabricated, the single-mode Y-junction, avoids major predesign considerations altogether, since it relies basically on symmetry alone for its operation.

1.3   LOSS MECHANISMS

Because of the inherent smoothness and uniformity of optical fibres, the principle loss mechanisms in fibre-based devices are associated with geometrical changes from uniformity. For example, bending loss limits the radius of ring resonators etc., and approximately adiabatic tapering limits the rate of drawdown of fused-taper couplers. However, losses due to bending and tapering are now well understood and quantified, and, consequently, these can be minimized.

In the case of compact planar waveguides and devices, there are four loss mechanisms which need to be addressed. These cover: (i) splicing between the planar waveguide and fibre; (ii) bend loss, because of the tight bends required in compact circuitry; (iii) scattering loss, due to the roughness introduced through the fabrication method; and (iv) substrate leakage, due to tunnelling of light from the lower-index guiding region to the higher-index substrate. As will be shown in later chapters, splice loss can, in theory, be made almost arbitrarily small, and bend loss, even from very tight bends, can be minimized by judicious choice of bend geometry and material parameters. Scattering loss due to surface roughness can also be reduced by decreasing the mean roughness amplitude, and by ensuring that the correlation length of the roughness lies outside the range of length scales associated with the coupling of guided light to radiation. Substrate leakage can be made arbitrarily small by ensuring a sufficiently thick buffer layer between the guiding region and the substrate.

1.4   WAVEGUIDE PROPAGATION AND DEVICE DESIGN

The thrust of subsequent chapters is to provide both physical insight and an appropriate range of analytical and numerical techniques for analysing and quantifying propagation along planar waveguides and through basic planar devices. Whilst for convenience and simplicity, the work is premised on single-mode buried channel waveguides (BCWs) and corresponding devices of square or rectangular cross-section withstep-index profiles, the techniques introduced are readily generalized and applied to planar waveguides and devices of quite arbitrary cross-sectional geometries and refractive-index profiles, and also to multimode planar devices.

The following chapter briefly summarises current techniques used for planar waveguide and device fabrication. Its main purpose is to provide background information to help the reader appreciate the physical origin of the propagation problems to be studied in subsequent chapters. Chapter 3 addresses basic modal propagation in BCWs, starting with Maxwell's equations, and then introduces the weak-guidance approximation in order to reduce the determination of the modal fields and propagation constants to a scalar problem. The subsequent Chapters 4–6 then introduce various approximation and numerical methods, and apply them to determine the modes of the basic step-profile, square-core BCW. Chapters 7–10 address and quantify each of the four basic loss mechanisms outlined in Section 1.3, while Chapter 11 delineates the design of optimally low-loss, tight bends. The remaining Chapters analyse propagation through basic BCW devices, covering various types of single-mode couplers in Chapter 12, large-angle single-mode X-junctions in Chapter 13 and symmetric, single-mode Y-junctions in Chapter 14.

Many of the analytical techniques used in this book are either drawn from, or adapted from, more general results presented elsewhere (Snyder and Love, 1983), so that the reader is encouraged to refer to this text if further details are required.

REFERENCES

Snyder, A. W. and Love, J. D. (1983) *Optical waveguide theory*. London: Chapman & Hall.

2
Fabrication techniques

This chapter provides a brief review of existing and emerging techniques for the fabrication of silica-based planar waveguides and devices. The review is not designed to be comprehensive, nor to cover detailed aspects of each technique, but is provided as a useful background for the physical understanding and analytical development of planar waveguides and devices in subsequent chapters. For convenience, the fabrication techniques discussed here can be categorized into three families, covering those processes which rely on: (i) the shaping of appropriate-index materials into waveguides and devices; (ii) changing the index by replacing some of the ions in the material with different ions; and (iii) changing the index of the material through its molecular response to particle or photonic beam bombardment.

2.1 DEPOSITION AND ETCHING TECHNIQUES

Processes based on deposition and etching techniques for silica-based glasses start with a crystalline planar substrate, typically a silicon wafer or optically flat silica glass, such as quartz. Silicon wafers have the advantages of a high degree of planarity, ready adhesion of deposited silica and, because of their excellent heat dissipation property, the potential for hybridization of optical and electronic components onto a common substrate. The larger size wafers are also cost effective, as they enable multiple copies of devices to be fabricated simultaneously on a single wafer. Before considering the different stages of the fabrication process, it is helpful to summarize the various deposition and etching techniques currently available.

2.1.1 Flame hydrolysis

Flame hydrolysis (FHD) (Kawashi, 1990) generates small glass particles by feeding a mixture of vapours, such as silicon tetrachloride ($SiCl_4$), titanium tetrachloride ($TiCl_4$) and germanium tetrachloride ($GeCl_4$) into an oxy-hydrogen torch. The resulting combustion process produces a white soot of fine glass particles which are deposited directly onto the substrate. The refractive index is determined by the relative flow rates of the constituent gases. Since flame hydrolysis produces a porous glass layer, the fabricated waveguide structure has to be consolidated by annealing, through heating, to around 1200°C or 1300°C for an extended period to consolidate the material and minimize propagation scattering loss.

2.1.2   Chemical vapour deposition

Chemical vapour deposition (CVD) and its variants, such as modified chemical vapour deposition (MCVD) and plasma–enhanced chemical vapour deposition (PECVD) (Giroult-Matlakowski *et al.*, 1994), can all produce thin layers of deposited silica or doped silica by using admixtures of either vapours or gases (Valette *et al.*, 1989). MCVD relies on thermal heating for the creation and deposition of the glass soot, whereas PECVD and its variants rely on ionic catalysts. One advantage of PECVD, over both FHD and other CVD processes, is that it is a low-temperature process, compatible with temperatures occurring in microelectronics processing. However, this potential advantage is negated if high-temperature annealing has to be invoked for smoothing out irregularities occurring during the deposition and etching processes. Another advantage that the PECVD process provides is the ability to monitor the refractive index of the deposited layers using *in-situ* ellipsometry (Charles *et al.*, 1993).

2.1.3   Sol-gel

It is possible to produce amorphous silica from the evaporation of silicate gel (Almeida, 1994) such as tetraethoxysilane (TEOS). A few drops of the gel are deposited on a silicon substrate that is spun to produce a uniform layer, whose thickness depends on the rotational speed, initial density and viscosity. The final state is obtained by evaporation after the film has attained a rigid gel state. Thick layers can be obtained by repeating this process. The waveguides can then be formed using etching techniques similar to those used in conjunction with FHD or CVD.

2.1.4   Plasma etching

Etching is the physico-chemical process whereby silica or doped silica is removed from a surface through bombardment by ions excited by a plasma, using, for example reactive ion etching (RIE). By covering part of a surface with a predesigned *mask*, or pattern, in order to inhibit etching, one can produce an arbitrarily complex layout of waveguides and devices. Typically, silica reacts strongly with elements like fluorine under ion bombardment, and a volatile compound is formed. Ions of the reactive element are created in the plasma, and an electric field is then used to direct the ions toward the areas to be etched.

2.2   FABRICATION BY DEPOSITION AND ETCHING

The starting point of the fabrication process is either a silicon wafer, some 5–20 cm in diameter and a few millimetres thick, or an optically flat surface, typically quartz, as shown schematically in Figure 2.1a. The silicon wafer is sometimes pre-heated in a furnace or immersed in a steam bath for an extended period, in order to oxidize a layer several microns thick within and below the surface, both to facilitate the adhesion of deposited silica and to form part of the buffer layer. Pure silica can then be deposited using FHD or PECVD.

In a typical fabrication process, a uniform buffer layer of pure silica is deposited on the surface of the wafer to a depth of the order of 10 μm or more, as shown schematically in Figure 2.1b. The choice of thickness, in terms of waveguiding, should be large enough to minimize the loss of light through tunnelling or leakage to the silicon substrate, which has a much higher index than silica. The precise depth of the buffer is determined by the maximum acceptable substrate leakage loss, and will be quantified in Chapter 9.

A second layer of glass is then deposited on the buffer layer to constitute the core layer in Figure 2.1c. The core index must be slightly higher than the substrate index in order to provide light guidance, and also to be compatible with the core-cladding index difference of standard telecommunications single-mode fibres which are spliced to the planar devices.

The thickness of the core layer is determined by the dual requirements that the waveguide be single-moded and that the splice loss to single-mode fibre be minimal. These aspects will be discussed in Chapters 5 and 7, respectively. The higher index of the core layer can be achieved either by doping with suitable elements, such as phosphorus, nitrogen, fluorine and germanium, as in the fibre MCVD process, or by using stoichiometric effects. In the latter, the ratio of oxygen to silane gases in the PECVD process is reduced so that effectively SiO_x is deposited, where $x < 2$.

Lithographic processes are then used to deposit a predesigned mask on the top of the core layer, whose size is shown in Figure 2.1d. In the example shown, the region immediately below the mask defines the core of a rectangular waveguide. The mask dimensions are determined by theoretical analysis for the particular layout of waveguides and devices required, and, using computer-aided design (CAD) techniques, the actual mask is fabricated to sub-micron accuracy.

Because of the relatively large size of the wafer and the relatively small area occupied by planar devices, a large number of identical or different device designs can be accommodated, e.g. a concatenation of the low-loss Y-junctions

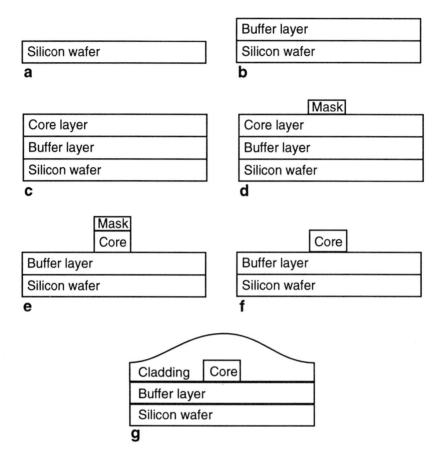

Figure 2.1 Fabrication steps for deposition/etching waveguide fabrication: (a) cross-section of a silicon wafer subtrate; (b) deposited substrate and buffer layer; (c) deposited bilayer of core and buffer; (d) cross-section of mask deposited on the core layer; (e) etching of the core layer; (f) removal of the mask; (g) deposited cladding layer and buried channel waveguide of rectangular cross-section.

of Chapter 14 for a 1 × 8 splitter occupies an area approximately 5 mm × 20 mm. With the mask *in situ*, the parts of the core layer not covered by the mask are etched away, as shown in Figure 2.1e.

The mask is then removed chemically to leave a rectangular cross-section, rib-shaped pattern on the top of the buffer layer, as shown in Figure 2.1f. A final deposit of silica completes the cladding surrounding the core to produce the buried channel waveguide shown in Figure 2.1g. The upper surface of the top layer tends to follow the contour of the buffer layer and rib waveguide

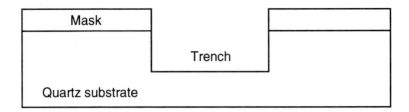

Figure 2.2 Trench etched in a quartz substrate to form the core region.

with a constant-thickness layer of similar thickness to the buffer layer. It is important to ensure that the index of the cladding matches that of the buffer as closely as possible, to facilitate the design and fabrication of devices, such as the single-mode couplers of Chapter 12.

2.2.1   Alternative approach

A variation on the procedure outlined above uses a glass substrate and one of the etching techniques to form a trench, as shown in Figure 2.2, using a mask which is the complement of that used in Figure 2.1e. The trench forms the guiding core region, which, after removal of the mask, is filled to the level of the surrounding quartz using one of the deposition techniques. A final layer of silica then forms the buried channel waveguide.

2.3   FABRICATION BY ION-EXCHANGE

The ion-exchange method relies on replacing the ions in a glass substrate with different ions by a diffusion technique in order to effect an increase in refractive index (Lerminiaux, 1992). This can be readily illustrated by the sodium–potassium process. Starting with an optically flat soda-lime glass, a complement mask is deposited on the surface, as shown in Figure 2.3. This mask is the reverse of the mask used in the deposition/etching processes, as the area not covered by the mask defines the cores of the waveguides and devices. The surface is then immersed in a liquid salt bath of potassium nitrate at high temperature for an extended period. Potassium ions migrate into the glass matrix and replace sodium ions, which in turn migrate out of the glass into the liquid. The net effect is to raise the refractive index.

The result of this process, after removal of the salt bath and mask, is to leave a waveguiding structure immediately below the air–glass interface. The profile is graded, because of the nature of the diffusion process, and the cross-section

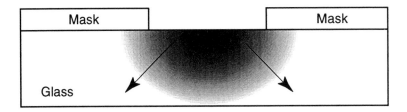

Figure 2.3 Schematic of the ion-exchange process.

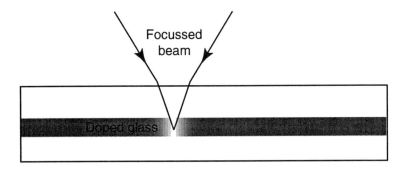

Figure 2.4 Schematic of direct-write processes.

of the waveguide is approximately semi-circular, with the highest index value at the interface. Because of its non-circular profile, such a waveguide would result in significant loss on splicing to a circular fibre. However, if, during the diffusion process, a strong electric field is applied across the glass block, the ions migrate much farther into the glass, resulting in a buried waveguide with a much more circular core cross-section.

2.4 DIRECT-WRITE PROCESSES

Planar waveguides can be fabricated directly in homogeneous or layered optical materials without the need for a mask by locally changing the index of the material through its molecular response to focussed particle (Townsend *et al.*, 1994) or photonic beam bombardment (Meltz and Morey, 1994), as shown schematically in Figure 2.4. A beam of protons can alter the local index through molecular bond modifications, so that by focussing and translating the beam it is possible to write buried waveguides of prescribed width and depth.

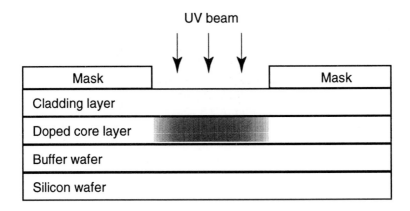

Figure 2.5 Schematic of the doped trilayer.

2.4.1   Photo-induced waveguides

It is possible to increase the refractive index of germania-doped silica optically by illumination with an intense UV beam (Meltz and Morey, 1994; Moss *et al.*, 1994). This technique, which has been used to write diffraction gratings directly into fibres, is now being used for the direct writing of buried waveguides and gratings, using trilayers of glass deposited onto a silicon substrate, as shown in Figure 2.5. The central slab layer consists of germania-doped silica, while the top and bottom cladding layers are essentially pure silica. A reverse mask, as used in the ion-exchange process, can be deposited on the top surface and a UV beam shone through it to produce the waveguide in the core layer. This procedure has been used successfully to write a single-mode coupler (Maxwell and Ainslie, 1995).

However, the mask can be dispensed with altogether and the waveguide written directly using an appropriately-focussed UV beam, which is translated along the length of the trilayer. By allowing both x- and y-movement of the beam, devices can also be written directly. This fabrication method has the potential to dramatically simplify the waveguide fabrication process, as it dispenses with both the masking and etching steps, and consequently avoids the problems associated with void formation when filling deep trenches or closely separated walls, such a those found in planar couplers and Y-junctions.

The trilayer illustrated in Figure 2.5 has a step-profile, defined by the buffer–core–cladding layers, which, therefore, constitutes a slab waveguide. After direct writing of the BCW, defined by the shaded core region, the slab waveguide still provides vertical guidance on either side of the BCW. This guidance could lead to interference within the trilayer. Light radiated from a BCW, or device, and trapped within the core layer could spread sufficiently far horizontally to affect propagation in a neighbouring BCW, or device. However, it is possible to minimize this effect by co-doping the core layer with germanium and an appropriate level of boron. The former ensures that the core remains photosensitive, and the latter ensures that the core refractive index now matches that of the pure silica buffer and cladding layers (Maxwell and Ainslie, 1995). Consequently, after UV-writing the BCW core in Figure 2.5, the surrounding buffer–core–cladding layers all have the same uniform refractive index (of silica), and any light lost from the BCW will generally radiate out in all directions.

2.5   PLANAR WAVEGUIDES AND DEVICES

The net result of the fabrication processes described above is to produce BCWs and devices with cross-sectional geometries and refractive index profiles defined by the particular techniques involved. Etching and deposition fabrication tends to produce waveguides and devices with nominally square or rectangular cores and step profiles, whereas ion-exchange fabrication results in graded profiles with no well-defined core–cladding boundary and cross-sectional geometries varying from circular to highly elliptical. Direct-write processes currently being developed have the potential to produce fairly arbitrary predetermined cross-sectional geometries and refractive index profiles as well.

2.5.1   Predesign

In each of these processes, it is necessary to predetermine all the parameters for a particular type of waveguide or device prior to fabrication, especially for mask design and fabrication. This in turn requires knowledge of the cross-sectional geometry and profile produced by the particular process. The main purpose of the following chapters is to show how to determine propagation and device characteristics from the geometry and profile information. In particular, the important aspects of substrate leakage, scattering loss, bend loss and splice loss will be quantified, since they determine the overall loss of any planar device and associated waveguides and spliced fibres.

Most silica-based planar waveguides and devices have only a small variation in refractive index over their cross-sections, so that it is sufficient to use a scalar analysis based on the weak-guidance approximation. This considerably simplifies the analysis, as we show in the next chapter.

REFERENCES

Almeida, R. M. (1994) Sol-gel silica films on silicon substrates. *International Journal of Optoelectronics*, **9**, 135–142.

Charles, C., Giroult-Matlakowski, G., Boswell, R. W., Goullet, A., Turban, G. and Cardinaud, C. (1993) Characterization of silicon dioxyde films deposited at low pressure and temperature in a helicon diffusion reactor. *Journal of Vacuum Science and Technology A*, **11**, 2954–2963.

Giroult-Matlakowski, G., Charles, C., Durandet, A., Boswell, R. W., Persing, H. M., Armand, S., Perry, A. J., Loyd, P. D. and Hyde, S. R. (1994) Deposition of silicon dioxide films using the helicon diffusion reactor for integrated optics applications. *Journal of Vacuum Science*, **2**, in press.

Kawashi, M. (1990) Silica waveguides on silicon and their application to integrated-optic components. *Optical and Quantum Electronics*, **22**, 391–416.

Lerminiaux, C. (1992). Glass planar devices made by ion exchange technique. Pages 262–263 of: *Proceedings OFC-92*.

Maxwell, G. D. and Ainslie, B. J. (1995) Demonstration of a directly written directional coupler using UV-induced photosensitivity in a planar silica waveguide. *Electronics Letters*, **31**, 95–96.

Meltz, G. and Morey, W. W. (1994) Photoinduced refractivity in germanosilicate waveguides. Pages PD6-2–PD6-4 of: *Integrated photonics research*.

Moss, D., Ibsen, M., Ouellette, F., Leech, P., Faith, M., Kemeny, P., Leistiko, O. and Poulsen, C. V. (1994) Photo-induced planar germanosilicate waveguides. Pages 333–336 of: *19th Australian conference on optical fibre technology*.

Townsend, P. D., Chandler, D. J. and Zhang, L. (1994) *Optical effects of ion implantation*. Cambridge: Cambridge University Press.

Valette, S., Renard, S., Denis, H., Jadot, J.P., Fournier, A., Philippe, P., Gidon, P., Grouillet, A. M. and Desgranges, E. (1989) Si-based integrated optics technologies. *Solid State Technologies*, **32**, 69–74.

3
Weak-guidance approximation

Maxwell's equations, or, equivalently, the vector wave equations, for the electri-cand magnetic fields are the starting point of any optical waveguiding analysis. However, the complexity of either set of equations, due to coupling of the three-scalar components of both the electric and magnetic field vectors, makes them very unwieldy, and as a result there are very few exact analytical solutions. Those that are available (Snyder and Love, 1983, Chapter 12) do not include the types of cross-sectional geometries and refractive-index profiles encountered inplanar waveguides.

Conveniently, silica-based planar waveguides normally have only a slight variation in refractive index over their cross-sections, partly because of the need to minimize splice loss with standard telecommunications single-mode fibres, and partly because of the nature of the fabrication processes, as outlined in the previous chapter. The slight variation in index is most helpful, as it enables the vector equations for the electromagnetic fields to be replaced by a single scalar equation involving only a single component of the electric field, with-negligible loss of accuracy. This simplification is known as the *weak-guidance approximation* (Snyder, 1969; Gloge, 1971), and although it is a purely scalar analysis, vector corrections for polarization effects can be incorporated as correction terms, as we will show in Section 3.6. A more thorough discussion of the weak-guidance approximation can be found elsewhere (Snyder and Love, 1983, Chapter 13).

3.1 WAVEGUIDE PARAMETERS

Waveguiding structures with the types of cross-sections illustrated in Figure 3.1 are all characterized by a central region of refractive index which is higher than that of the surrounding. This region – the *core* – together with the surrounding region of normally uniform index – the *cladding* – ensure that light is guided along the structure. In the case of the step profile, the core has a uniform index. It may be helpful to think of guidance due to this profile as a sequence of reflections of ray paths from the interface between the core and the cladding, i.e. *total internal reflection*, although for single-mode waveguides this is not an accurate picture for the purposes of quantification because of diffraction,

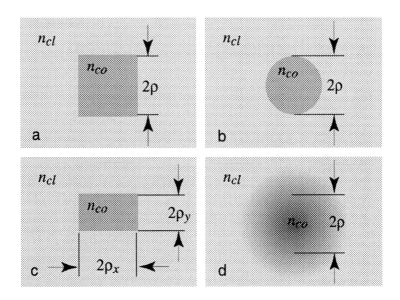

Figure 3.1 Four typical waveguide cross-sections: (a) step-index square-core buried channel waveguide (BCW); (b) step-index fibre; (c) rectangular core with the characteristic dimension defined by the geometric mean $\rho = (\rho_x \rho_y)^{1/2}$; (d) diffused waveguide with a graded profile with the characteristic dimension ρ defined by the e-folding distance of its index profile $n(x, y)$.

i.e. wavelength-dependent effects. The propagation characteristics of all these waveguides can be expressed in terms of the following parameters.

3.1.1   Characteristic dimension

The characteristic dimension of the waveguide cross-section is the linear dimension ρ that quantifies the transverse variation in index of the waveguiding structure. With reference to the step-profile waveguides in Figure 3.1, ρ corresponds to the core radius of the fibre and to the half-side of the square-core waveguide. For the rectangular core waveguide, the geometric mean of the core half-side lengths is used. For waveguides with no well-defined core–cladding interface, such as the graded-index waveguides produced by the ion-exchange process, ρ can be defined as the e-folding distance of the variable part of the profile.

3.1.2   Relative index difference

If we define n_{co} and n_{cl} as the representative core index and uniform cladding index of the waveguide, respectively, the *relative index difference* Δ is defined as

$$\Delta = \frac{n_{co}^2 - n_{cl}^2}{2n_{co}^2} \tag{3.1}$$

If the structure has a graded profile, formed in the ion-exchange process, for example, then n_{co} and n_{cl} are taken to be, respectively, the maximum and minimum index values of the profile $n(x,y)$. For weakly-guiding waveguides with $n_{co} \approx n_{cl}$, then $\Delta \ll 1$ and the above expression can be accurately approximated by

$$\Delta \approx \frac{n_{co} - n_{cl}}{n_{co}}$$

and hence the alternative description of Δ as the *relative profile height*.

3.1.3   Waveguide parameter

The characteristic dimension ρ and the profile height Δ can be combined together with the operating wavelength λ of the source of excitation to form the dimensionless quantity V, known as the *degree of guidance* or *waveguide parameter*, defined as

$$V = \frac{2\pi}{\lambda}\rho n_{co}\sqrt{2\Delta} = \frac{2\pi}{\lambda}\rho(n_{co}^2 - n_{cl}^2)^{1/2} \tag{3.2}$$

with the quantity $(n_{co}^2 - n_{cl}^2)^{1/2}$ usually referred to as the *numerical aperture*. The degree of guidance V and the profile height Δ form the basis of the perturbation analysis that leads from Maxwell's equations to the scalar wave equation through the weak-guidance approximation.

3.1.4   Normalized index profile

It is sometimes convenient to represent the index profile $n(x,y)$ in a normalized form involving the profile height Δ. Accordingly, we define the *normalized index profile* $f(x,y)$ through the relationship

$$n^2(x,y) = n_{co}^2[1 - 2\Delta f(x,y)] \tag{3.3}$$

Using this definition and the fact that n_{co} and n_{cl} are the extreme values of the refractive index distribution, then $0 \le f(x,y) \le 1$, provided there are no depressions in the profile, as is normally the case with planar waveguides. For example, the step index profile of Figures 3.1a, b and c has $f = 0$ in the core and $f = 1$ throughout the cladding.

3.2   MODES AND FIELDS

The determination of propagation along optical waveguides requires knowledge of the refractive index profile resulting from any of the fabrication processes described in the previous chapter. We assume, for simplicity, that these waveguides are isotropic and non-absorbing, so that $n(x, y, z)$ is a real, scalar function. Cartesian axes are orientated such that the z-axis is parallel with the waveguide axis and, therefore, with the direction of propagation, and the x- and y-axes are in the waveguide cross-section. Waveguides are assumed to be translationally invariant, i.e. the profile does not change along their length, so that the index profile has only transverse dependence, i.e. $n = n(x, y)$.

The dependence of the refractive index on only the transverse coordinates allows the electric field $\mathbf{E}$ and magnetic field $\mathbf{H}$ vectors to be written in the separable forms

$$\mathbf{E}(x, y, z) = \mathbf{e}(x, y) \exp(i\beta z) \exp(-i\omega t) \tag{3.4a}$$

$$\mathbf{H}(x, y, z) = \mathbf{h}(x, y) \exp(i\beta z) \exp(-i\omega t) \tag{3.4b}$$

where $\mathbf{e}$ and $\mathbf{h}$ are vector expressions which contain, respectively, the transverse dependence of the electric and magnetic fields. The term βz in the first exponent denotes the accumulated phase change at distance z along the waveguide in terms of the propagation constant β, and the term $-i\omega t$ in the second exponent denotes the monochromatic time dependence of the phase in terms of the source frequency ω and time t. From here on, the time dependence is taken to be implicit in all field expressions, where appropriate.

For subsequent discussion, it is convenient to distinguish between the longitudinal components of the electric and magnetic fields, denoted by the scalar quantities $e_z(x, y)$ and $h_z(x, y)$, respectively, and the corresponding transverse components, denoted by the vector quantities $\mathbf{e}_\perp(x, y)$ and $\mathbf{h}_\perp(x, y)$, respectively, where

$$\mathbf{e}(x, y) = \mathbf{e}_\perp(x, y) + e_z(x, y)\hat{\mathbf{z}} \tag{3.5a}$$

$$\mathbf{h}(x, y) = \mathbf{h}_\perp(x, y) + h_z(x, y)\hat{\mathbf{z}} \tag{3.5b}$$

and $\hat{\mathbf{z}}$ is the unit vector parallel to the z axis.

3.2.1   Bound modes

The determination of the transverse field dependencies $\mathbf{e}$ and $\mathbf{h}$, together with the propagation constant β, constitutes an eigenvalue problem, in as much as the solutions of Maxwell's equations must satisfy boundary and boundedness conditions. This problem can be likened to the determination of the mechanical vibrations, or resonances of a two-dimensional membrane. Accordingly, there are certain discrete electromagnetic resonances in the cross-section, corresponding to *bound modes* of the waveguide.

Each bound mode has a discrete value of the propagation constant β, which is a solution of the eigenvalue equation, and an associated electric and magnetic field distribution which remains invariant along the length of the waveguide. The flow of electromagnetic power in a mode is always parallel to the waveguide axis. Depending on the parameter values and source wavelength for a particular waveguide, one or more modes can be supported. Although the complete guided field of the waveguide comprises a linear superposition of bound modes, we shall be primarily concerned with waveguides which support only one bound mode, the *fundamental* mode. These are referred to as *single-mode* waveguides.

The range of values that the propagation constant β of any bound mode can take is related to the maximum and minimum core index values. Thus, if n_{co} is the maximum value and n_{cl} the minimum value, i.e. the uniform cladding index, then (Snyder and Love, 1983)

$$kn_{cl} \leq \beta \leq kn_{co} \tag{3.6}$$

where $k = 2\pi/\lambda$ is the wavenumber and λ is the source wavelength. The *cutoff* of a mode corresponds to the smallest permissible value of the propagation constant, i.e. $\beta = kn_{cl}$. Below this value, the mode is no longer guided. In practive, the *cutoff wavelength* of the second mode is used to define the largest wavelength below which the BCW is single moded.

3.3   VECTOR WAVE EQUATION

Maxwell's equations link the spatial and temporal dependence of the vector electric and magnetic fields $\mathbf{E}$ and $\mathbf{H}$, and have the following forms for source-free, dielectric media (Snyder and Love, 1983, Equation 30–1)

$$\nabla \times \mathbf{E} = ik(\mu_0/\epsilon_0)^{1/2}\mathbf{H} \tag{3.7a}$$
$$\nabla \times \mathbf{H} = -ikn^2(\mu_0/\epsilon_0)^{1/2}\mathbf{E} \tag{3.7b}$$

where ϵ_0 and μ_0 are, respectively, the free space dielectric constant and permeability, n is the refractive index profile and $k = 2\pi/\lambda$ is the free-space wavenumber in terms of the source wavelength λ. By taking the curl of (3.7a) and using (3.7b) to eliminate the magnetic field, we find

$$\nabla \times \nabla \times \mathbf{E} - k^2 n^2 \mathbf{E} = 0 \tag{3.8}$$

which is the vector wave equation for the electric field. This equation can be recast into a form more suitable for waveguiding problems by substituting the modal definition (3.4a) into it, and thus obtaining

$$(\nabla_\perp^2 + k^2 n^2(x,y) - \beta^2)\mathbf{e} = -(\nabla_\perp + i\beta\hat{\mathbf{z}})\mathbf{e} \cdot \nabla_\perp \ln n^2(x,y) \tag{3.9}$$

where the divergence condition $\nabla(n^2\mathbf{E}) = 0$ has been used, and $\nabla_\perp^2$ is the transverse vector Laplacian operator. The next step is to decompose the field e into its transverse and longitudinal components using (3.5a) to obtain the equation describing the dependence of the transverse components

$$(\nabla_\perp^2 + k^2 n^2(x,y) - \beta^2)\mathbf{e}_\perp = -\nabla_\perp[\mathbf{e}_\perp \cdot \nabla_\perp \ln n^2(x,y)] \qquad (3.10)$$

together with a similar expression for the longitudinal z-component e_z. It is important to notice that (3.10) does not involve the longitudinal component. The latter can be deduced directly from the transverse components through the expression (Snyder and Love, 1983, Section 30–1)

$$e_z = \frac{i}{\beta}\left[\nabla_\perp \cdot \mathbf{e}_\perp + (\mathbf{e}_\perp \cdot \nabla_\perp)\ln n^2(x,y)\right] \qquad (3.11)$$

as can be seen by inserting (3.4) and (3.5) into Faraday's law (3.7b) and using (3.7a) to eliminate the magnetic field.

However, even using the simpler form (3.10), it is not possible to solve analytically for the transverse component of the vector electric field for the types of profiles and cross-sections encountered in planar waveguides, and even a numerical solution is complicated because of the implicit coupling of the components of the field in this equation. Fortunately, the slight variation in profile encountered in practical silica-based waveguides enables a dramatic simplification to be introduced.

3.4   WEAK-GUIDANCE APPROXIMATION

When the profile height of a waveguide becomes vanishingly small, i.e. $\Delta \to 0$ and thus the core and cladding indices become very similar, it is intuitive that the guided field of a bound mode should resemble a plane wave in a bulk uniform medium propagating parallel to the waveguide axis. Since such plane waves are purely transverse electromagnetic, we anticipate that the longitudinal field components will be small compared with the transverse components when $\Delta \ll 1$. Moreover, since, for the step profile, total internal reflection can occur for arbitrarily small Δ, as long as the reflection angle is small enough, field confinement is possible in the limit $\Delta \to 0$. Accordingly, we anticipate modal solutions of the wave equation having the properties of (inhomogeneous) plane waves.

3.4.1   Perturbation analysis

In order to perform a formal perturbation expansion of the transverse vector wave equation (3.10), it is convenient to re-express this equation in terms of the normalized index profile $f(x,y)$, profile height Δ, degree of guidance V and the characteristic dimension ρ of Section 3.1 as

$$(\rho^2 \nabla_\perp^2 + U^2 - V^2 f)\mathbf{e}_\perp = -\rho \nabla_\perp (\mathbf{e}_\perp \cdot \rho \nabla_\perp \ln n^2) \tag{3.12}$$

We have introduced the normalized core mode parameter U which is related to the propagation constant by

$$U = \rho \left(k^2 n_{\text{co}}^2 - \beta^2\right)^{1/2} \tag{3.13}$$

If we consider Δ as the perturbation parameter, then we can expand the transverse field components and normalized propagation constant in a power series in Δ

$$\mathbf{e}_\perp(\Delta) = \tilde{\mathbf{e}}_\perp + \Delta \mathbf{e}_\perp^{(1)} + \Delta^2 \mathbf{e}_\perp^{(2)} + \cdots \tag{3.14a}$$

$$U(\Delta) = \tilde{U} + \Delta U^{(1)} + \Delta^2 U^{(2)} + \cdots \tag{3.14b}$$

where the superscript indicates the order of the perturbation and the tilde denotes the zeroth-order solution. The right hand side of the normalized vector wave equation (3.12) is also expanded for small Δ using (3.3). The expansion of the logarithmic term for small argument is

$$\nabla_\perp \ln n^2 = -2\Delta \nabla_\perp f - 4\Delta^2 \nabla_\perp f^2 + \cdots \tag{3.14c}$$

3.4.2   Scalar wave equation

Substituting the expansions (3.14) into the normalized vector wave equation (3.12) and equating terms of the same order in Δ leads to the zeroth-order field or scalar wave equation

$$(\rho^2 \nabla_\perp^2 + \tilde{U}^2 - V^2 f)\tilde{\mathbf{e}}_\perp = 0 \tag{3.15}$$

where the zeroth-order approximation $\tilde{\beta}$ of the propagation constant β is related to $\tilde{U}$ through

$$\tilde{U} = \rho \left(k^2 n_{\text{co}}^2 - \tilde{\beta}^2\right)^{1/2} \tag{3.16}$$

Finally, in cartesian coordinates the two scalar components of $\tilde{\mathbf{e}}_\perp$ decouple so they each obey the same scalar wave equation

$$(\nabla_\perp^2 + k^2 n^2(x,y) - \tilde{\beta}^2)\phi(x,y) = 0 \tag{3.17}$$

where $\phi(x,y)$ stands for either $\tilde{e}_x(x,y)$ or $\tilde{e}_y(x,y)$.

Thus we have shown that, to zeroth order in Δ, both transverse electric field components obey the same scalar wave equation (3.17). The boundary conditions associated with this equation require that $\phi(x,y)$ and all its first derivatives be continuous everywhere. The solutions of this equation must also be bounded everywhere and vanish at infinite distance from the core. These requirements constitute an eigenvalue problem for the modal fields and propagation constants.

If the scalar electric field components are substituted into (3.11), it is straightforward to show, that to zeroth order, the longitudinal field e_z vanishes, and that its lowest-order contribution is of order $\Delta^{1/2}$, given by

$$e_z(x, y) = i\frac{(2\Delta)^{1/2}}{V}(\rho\nabla_\perp \cdot \tilde{\mathbf{e}}_\perp) \tag{3.18}$$

3.4.3   Zeroth-order approximation

The scalar wave equation (3.17) was derived as the zeroth order approximation to the transverse vector wave equation in the limit $\Delta \to 0$. This approximation involves the index profile through the term $V^2 f(x, y)$ in (3.15), which implies that, in the limit $\Delta \to 0$, V should be kept fixed in order to ensure guidance. This would require setting $\lambda = 0$ when $\Delta = 0$ in order to keep V finite and arbitrary. However, for practical situations where Δ is small but finite and ρ/λ is large enough to ensure that V is non-zero, the zeroth order or weak guidance result is an excellent approximation to the full vector solution (Snyder and Love, 1983, Chapter 12).

3.4.4   Paraxial approximation

In a geometric optics analysis of waveguides (Snyder and Love, 1983, Part I), the paraxial approximation assumes that guided rays have paths whose directions are close to parallel to the waveguide axis, corresponding to a refractive index profile with only a slight variation in index. Since this slight variation in profile is also the basis of the weak-guidance approximation, the two approximations are intimately linked.

This can be demonstrated by examining the z-dependent phase term, $\exp(i\beta z)$, in the modal fields of (3.4). If we interpret the propagation constant, β, as the z-component of the local wave vector, then, in a uniform core of index n_{co}, we must have

$$\beta = kn_{co}\cos(\theta_z) \tag{3.19}$$

where k is the wavenumber and θ_z the angle between the local wave vector and the direction of propagation. Within weak-guidance, β is bounded by the limits in (3.6), and as $n_{co} \approx n_{cl}$, this implies that $\cos(\theta_z) \approx 1$, and hence θ_z must be small, in keeping with the paraxial approximation.

3.5   POLARIZATION OF THE MODAL FIELD

In the weak-guidance approximation, the lowest-order solution for the fundamental modal electric field is plane-polarized over the infinite waveguide cross-section. The scalar wave equation does not specify the direction of polarization and can only determine the spatial distribution of the e_x or e_y field components. Normally, there are two possible orthogonal polarization directions for a mode determined by the cross-sectional profile and geometry. Usually these directions are obvious from symmetry, and it is then convenient to align the x- and y-axes parallel to them. Thus one polarization of the mode has only the e_x component, yielding the x-polarized modal field while the other polarization has only the e_y component, giving the y-polarized modal field. These two scalar solutions have the same spatial field distribution so only their polarizations differ.

For the square-core waveguide which has the four-fold symmetry of Figure 3.2a, the orthogonal x- and y-polarized modal fields are degenerate since they have the same propagation constant β. In other words, any pair of orthogonal directions can be used with the same field distribution and propagation constant.

When the symmetry is less than four-fold, illustrated by the rectangular-core waveguide and the planar coupler illustrated, respectively, in Figure 3.2b and Figure 3.2c, the two orthogonal polarization states of the fundamental mode have the same field and scalar propagation constant. If we allow for higher-order corrections, the propagation constants then differ slightly and the structure is said to be *birefringent*. The respective polarization corrections to the scalar propagation constant, as calculated from the scalar wave equation, are based on a simple perturbation expression, which will be derived from the vector wave equation in the next section.

3.6   POLARIZATION CORRECTIONS

A simple perturbation correction to the scalar propagation constant $\tilde{\beta}$, which accounts for polarization effects, can be derived by combining the scalar wave equation (3.17) with the corresponding vector wave equation (3.10)

$$(\nabla_\perp^2 + k^2 n^2(x,y) - \beta^2)\mathbf{e}_\perp = -\nabla_\perp[\mathbf{e}_\perp \cdot \nabla_\perp \ln n^2(x,y)] \qquad (3.20a)$$
$$(\nabla_\perp^2 + k^2 n^2(x,y) - \tilde{\beta}^2)\tilde{\mathbf{e}}_\perp = 0 \qquad (3.20b)$$

If we multiply (3.20a) by $\tilde{\mathbf{e}}_\perp$ and (3.20b) by $\mathbf{e}_\perp$, subtract and integrate over the infinite cross-section A_∞ we get (Snyder and Young, 1978)

$$\beta^2 - \tilde{\beta}^2 = \frac{\int_{A_\infty}(\nabla \cdot \tilde{\mathbf{e}}_\perp)\mathbf{e}_\perp \cdot \nabla_\perp \ln n^2 \mathrm{d}A}{\int_{A_\infty}\mathbf{e}_\perp \cdot \tilde{\mathbf{e}}_\perp \mathrm{d}A} \qquad (3.21)$$

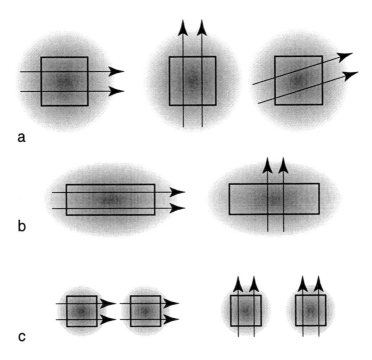

Figure 3.2 Fundamental modes of three waveguiding structures: (a) the square-core step-index waveguide with x- and y-polarized fundamental modes. Due to degeneracy the fundamental mode can have arbitrary plane polarization; (b) the rectangular-core waveguide and (c) the square-core coupler with supermodes polarized parallel to either of the symmetry axes (see Chapter 12).

This expression is exact and it holds for any profile, step or graded. We define the polarization correction $\delta\beta$ to the scalar, or weak-guidance propagation constant by

$$\beta = \tilde{\beta} + \delta\beta \tag{3.22}$$

and factorize the left hand side of (3.21)

$$\beta^2 - \tilde{\beta}^2 = \delta\beta(\beta + \tilde{\beta})$$

The bound-mode propagation constant β satisfies $kn_{cl} \leq \beta \leq kn_{co}$ which, when combined with the weak-guidance condition $n_{co} \approx n_{cl}$, enables the sum $\beta + \tilde{\beta}$

to be well approximated by $2kn_{co}$. Thus, in terms of V and Δ

$$\beta^2 - \tilde{\beta}^2 \approx \frac{V}{\rho} \left(\frac{2}{\Delta}\right)^{1/2} \delta\beta$$

Finally, on replacing $\mathbf{e}_\perp$ by the weak-guidance approximation $\tilde{\mathbf{e}}_\perp$ in (3.21), we obtain

$$\delta\beta = \frac{(2\Delta)^{1/2}}{2\rho V} \frac{\int_{A_\infty} (\rho\nabla_\perp \cdot \tilde{\mathbf{e}}_\perp)\tilde{\mathbf{e}}_\perp \cdot \rho\nabla_\perp f \, \mathrm{d}A}{\int_{A_\infty} \tilde{\mathbf{e}}_\perp^2 \, \mathrm{d}A} \tag{3.23}$$

for the polarization correction in terms of weak-guidance quantities.

3.7 POWER ATTENUATION

In subsequent chapters, we will be addressing various physical mechanisms which give rise to a loss of power from the fundamental mode. The attenuation of a mode's guided power can be expressed in terms of a power attenuation, or loss coefficient γ. If γ is independent of position along the waveguide, the modal power, $P(z)$, at distance z along the waveguide is related to the initial power, $P(0)$, by

$$P(z) = P(0)\exp(-\gamma z) \tag{3.24}$$

In the waveguide cross-section, the flow of power at each position z is parallel to the waveguide axis and the power density is given by the time averaged Poynting vector. Within weak-guidance, the Poynting vector is expressible in terms of the electric field according to (Snyder and Love, 1983, Chapter 13)

$$\frac{n_{co}}{2} \left(\frac{\epsilon_0}{\mu_0}\right)^{1/2} |\tilde{\mathbf{e}}_\perp|^2 \tag{3.25}$$

and the total power in the mode is determined by integrating the Poynting vector over the infinite cross-section of the waveguide.

3.8 LOCAL MODES

The concept of a local mode and its analytical representation is well established (Snyder and Love, 1983, Chapter 19). As we showed in Section 3.2, a mode of a straight, uniform waveguide propagates unattenuated along its length. If the waveguide now departs from straightness or uniformity, due to, e.g. tapering or bending, power is necessarily lost to radiation. However, if the departure from straightness or uniformity is sufficiently slow, it is intuitive that the modal field and propagation constant will follow the change in cross-sectional geometry, and the power of the mode will be approximately conserved.

The first result states that the modal field and propagation constant are determined by the local cross-section at each position along the waveguide. Accordingly, the accumulated phase along the nonuniform waveguide requires an integration over the local propagation constant. The second result is equivalent to requiring that the modal field be approximately orthonormal. Putting these requirements together, the scalar local mode has the following form:

$$\mathbf{E} = \hat{\mathbf{e}}(x, y, \beta(z)) \exp\left(i \int_0^z \beta(z)\,\mathrm{d}z\right) \tag{3.26}$$

where the 'hat' denotes orthonormality, i.e. unit normalization, and the integration denotes the accumulated phase. Note that both the field, $\hat{\mathbf{e}}(x, y, \beta(z))$, and propagation constant, $\beta(z)$, are z-dependent.

3.9 SUPERMODES

In planar waveguides or devices where there is more than one core present, e.g. the twin parallel cores of the couplers to be discussed in Chapter 12, the term 'supermode' has evolved to denote a normal mode of the complete multicore waveguide. Part of the reason for the introduction of this nomenclature is to distinguish these modes from the modes associated with each core and its surrounding cladding. Whilst a mode of a single core and cladding is not generally a mode of the composite multicore waveguide, propagation along the multicore waveguide can be analysed in terms of coupling between modes of different cores, using coupled mode theory, as will be described in more detail in Section 12.4. The equivalent analysis in terms of a superposition of supermodes will be described in Section 12.5.

For a multicore waveguide with cores which vary sufficiently slowly along its length, we can readily extend the local mode concept of Section 3.8 to apply to local supermodes, as will be used in Section 12.7.2.

REFERENCES

Gloge, D. (1971) Weakly guiding fibers. *Applied Optics*, **10**, 2252–2258.

Snyder, A. W. (1969) Asymptotic expressions for eigenfunctions and eigenvalues of dielectric or optical fibers. *IEEE Transactions on Microwave Theory and Techniques*, **MTT–17**, 1130–1138.

Snyder, A. W. and Love, J. D. (1983) *Optical waveguide theory*. London: Chapman & Hall.

Snyder, A. W. and Young, W. R. (1978) Modes of optical waveguides. *Journal of the Optical Society of America*, **68**, 297–309.

4
Approximation methods for modal analysis

Even within the weak-guidance approximation, there are no exact analytical solutions available for the modes of the waveguide profiles encountered in the fabrication processes described in Chapter 2. Accordingly, it is necessary to resort to the numerical methods that will be described in detail in Chapter 5 in order to obtain sufficiently accurate solutions. Nevertheless, it is possible to generate approximate analytical solutions for certain profiles and geometries. Such solutions often have an advantage over purely numerical solutions in that they reveal useful parametric dependencies.

In this chapter, we investigate various approximate solutions of the scalar wave equation for square- and rectangular-core step-profile waveguides based on trial functions for the scalar field, together with boundary and boundedness conditions. Depending on the technique employed, the matching condition either requires that the boundary conditions at the core–cladding interface be satisfied (as in the perfectly conducting limit of Section 4.1 and Marcatili's approximation of Section 4.2), or that a minimization procedure be imposed (as in the case of the Gaussian approximation of Section 4.4). The results obtained from these approximation methods, together with those obtained numerically in the following Chapter 5, will be presented and compared in Chapter 6.

4.1 PERFECTLY CONDUCTING LIMIT

When the waveguide V-value becomes very large, the modal field of the step-profile waveguide is almost totally confined to the core, with negligible field in the cladding. In other words there is virtually zero field on the core–cladding interface, and the core can be assumed to be surrounded by a zero-field metallic region. If we recall the definition

$$V = k\rho n_{\text{co}}\sqrt{2\Delta} \tag{4.1}$$

we see that the *perfectly conducting limit*, i.e. $V \to \infty$, is equivalent to either increasing the profile height Δ, or increasing the waveguide dimension ρ, or by lowering the wavelength so that the wavenumber $k = 2\pi/\lambda$ increases.

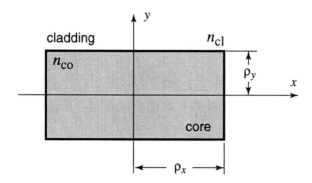

Figure 4.1 Axes for describing the rectangular-core, step-profile waveguide.

In this limit, there is a simple separable solution of the scalar wave equation for a uniform square or rectangular core surrounded by such a boundary. Relative to the cartesian axes of Figure 4.1, the scalar wave equation within the core has the form

$$\left(\frac{\partial^2}{\partial x^2} + \frac{\partial^2}{\partial y^2} + k^2 n_{co}^2 - \beta^2 \right) \psi(x, y) = 0 \tag{4.2}$$

where β is the propagation constant and $\psi(x, y)$ the transverse spatial dependence of the scalar fundamental-mode field. Thus, the fundamental-mode solution of this equation, i.e. the solution with the largest value of propagation constant and satisfying $\psi(x, y) = 0$ on the four boundaries $x = \pm \rho_x$ and $y = \pm \rho_y$, is readily found from separation of variables to have the form

$$\psi(x, y) = A \cos \frac{\pi}{2} X \cos \frac{\pi}{2} Y \tag{4.3}$$

where $X = x/\rho_x$, $Y = y/\rho_y$ and A is a constant. The eigenvalue equation follows from substituting (4.3) into (4.2), and can be expressed as

$$U = \frac{\pi}{2} \left(\frac{\rho_x}{\rho_y} + \frac{\rho_y}{\rho_x} \right)^{1/2} \tag{4.4}$$

where the modal parameter U is defined as

$$U = \left[\rho_x \rho_y \left(k^2 n_{co}^2 - \beta^2 \right) \right]^{1/2} \tag{4.5}$$

In the special case of the square, $\rho_x = \rho_y = \rho$, (4.4) reduces to

$$U = \frac{\pi}{\sqrt{2}} \tag{4.6}$$

and U is now defined by

$$U = \rho\left(k^2 n_{\text{co}}^2 - \beta^2\right)^{1/2} \tag{4.7}$$

These analytical results also provide a check on the large-V limit of the results presented in the following sections.

4.2 MARCATILI APPROXIMATION

A well-known approximation method for solving the scalar wave equation for rectangular or square cross-sections with a step-index profile is due to Marcatili (Marcatili, 1969). The accuracy of this method increases with increasing values of V, but, nevertheless, it provides reasonable estimates for the range of V-values in the practical single-mode régime. The idea behind the Marcatili approximation is a logical extension of the perfectly-conducting solution discussed above.

Suppose we replace the rectangular dielectric core by the orthogonal intersection of two step-profile slab waveguides with claddings and cores of thicknesses $2\rho_x$ and $2\rho_y$, respectively, and corresponding indices n_{co} and n_{cl}. The geometry and refractive-index distribution are as shown in Figure 4.2. If the V-value for each slab waveguide is large, then the fundamental-mode field will be virtually confined to the core of each waveguide. This means that over the square or rectangular intersection of the two waveguides, i.e. the BCW core, the field will be strongly confined in both the x- and y-directions. In other words, we recover the solution given by (4.3) in the limit when both V-values for the intersecting waveguides are unbounded.

4.2.1 Analytical solution

The solution of the scalar wave equation in the core for the fundamental mode within the Marcatili approximation has the form

$$\psi(x, y) = A \cos U_x X \cos U_y Y \tag{4.8}$$

where the intersecting slab waveguides have core modal parameters defined by

$$U_x = \rho_x \left(k^2 n_{\text{co}}^2 - \beta_x^2\right)^{1/2}, \quad U_y = \rho_y \left(k^2 n_{\text{co}}^2 - \beta_y^2\right)^{1/2} \tag{4.9}$$

for the x- and y-orientated waveguides, respectively. Substituting (4.8) into (4.2) and recalling (4.5) leads to

$$U = \left(\frac{V_y}{V_x} U_x^2 + \frac{V_x}{V_y} U_y^2\right)^{1/2} \tag{4.10}$$

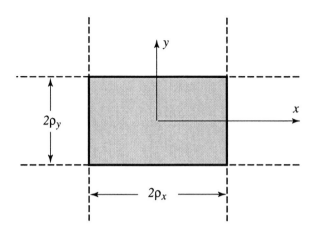

Figure 4.2 Orthogonal intersection of two planar waveguides forming an approximationof the rectangular-core buried channel waveguide.

for U in terms of U_x and U_y. The values of U_x and U_y are obtained from the eigenvalue equations for the x- and y-orientated planar waveguides, i.e. they are the solutions of (Snyder and Love, 1983, Chapter 12)

$$W_x = U_x \tan U_x, \quad W_y = U_y \tan U_y \tag{4.11}$$

respectively, having the smallest values of U_x and U_y where

$$W_x = \left(V_x^2 - U_x^2\right)^{1/2}, \quad W_y = \left(V_y^2 - U_y^2\right)^{1/2} \tag{4.12}$$

$$V_x = k\rho_x \left(n_{co}^2 - n_{cl}^2\right)^{1/2}, \quad V_y = k\rho_y \left(n_{co}^2 - n_{cl}^2\right)^{1/2} \tag{4.13}$$

In the special case of the square, $U_x = U_y = U_{sl}$, and(4.10) reduces to

$$U = \sqrt{2}U_{sl} \tag{4.14}$$

i.e. U is a constant multiple of the slab-waveguide U-value U_{sl}. As Marcatili's approximation is an extension of the perfectly conducting solution of Section 4.1, its accuracy decreases as $V \to 0$. Furthermore it predicts a spurious cutoff wavelength for the fundamental mode at $V = \pi/\sqrt{8} \approx 1.1107$ in the case of the square waveguide.

For both the rectangular- and square-core waveguides, the corresponding slab-waveguide eigenvalue equations are transcendental and must be solved numerically for the exact solutions for U_x and U_y. However, if we recall that the Marcatili approximation is more accurate for larger V-values, it is possible to obtain an explicit solution.

4.2.2   Explicit solution

As V_x, $V_y \to \infty$, both U_x, $U_y \to \pi/2$, enabling us to generate explicit, asymptotic solutions of the transcendental eigenvalue equations (4.11), namely (Snyder and Love, 1983, Chapter 12)

$$U_x = \frac{\pi}{2} \frac{V_x}{V_x + 1}, \quad U_y = \frac{\pi}{2} \frac{V_y}{V_y + 1} \tag{4.15}$$

On substituting (4.15) into (4.10), the explicit solution for the rectangular-core waveguide is

$$U = \frac{\pi}{2} (V_x V_y)^{1/2} \left(\frac{1}{(V_x + 1)^2} + \frac{1}{(V_y + 1)^2} \right)^{1/2} \tag{4.16}$$

which reduces to

$$U = \frac{\pi}{\sqrt{2}} \frac{V}{V + 1} \tag{4.17}$$

for the square-core waveguide. The two expressions reduce to (4.4) and (4.6), respectively, as V_x, $V_y \to \infty$, where V is given by (4.1) for the square-core waveguide.

4.3   EFFECTIVE INDEX METHOD

The Marcatili approximation discussed above decouples the square- or rectangular-core two-dimensional problem into two one-dimensional problems that can be solved independently. Comparison of Marcatili's approach with the perfectly conducting limit shows that the former is asymptotically exact in the limit $V \to \infty$. As indicated in Section 4.2, the Marcatili approximation becomes less accurate as V decreases and the modal field spreads farther and farther into the cladding.

One way to improve on the Marcatili approximation is to *couple* the two dimensions of the problem by introducing an *effective index* n^{eff} (Hocker and Burns, 1977; Chiang, 1986; Buus, 1984). This involves first solving (4.11) for the propagation constant in one of the dimensions, say x, as if the waveguide were a slab, and then determining an effective index

$$n_x^{\text{eff}} = \beta/k \tag{4.18}$$

This is used to construct an effective V-value for the y-direction by setting

$$V_y^{\text{eff}} = k\rho_y \sqrt{(n_x^{\text{eff}})^2 - n_{\text{cl}}^2} = V_y \frac{W_x}{V_x} \tag{4.19}$$

leading to the eigenvalue equation

$$W_y^{\text{eff}} = U_y^{\text{eff}} \tan U_y^{\text{eff}}, \quad W_y^{\text{eff}} = \left[(V_y^{\text{eff}})^2 - (U_y^{\text{eff}})^2 \right]^{1/2} \tag{4.20}$$

which is solved numerically. The final U-parameter is obtained from (4.10) where V_y and U_y are replaced by V_y^{eff} and U_y^{eff}, respectively.

The effective index method avoids the spurious cutoff problem of the Marcatili approach, but suffers from the same deficiency for determining the modal fields, i.e. the intersection of the fields of two slab waveguides is not a good approximation to the exact fields, especially near to the corners of the core–cladding interface.

4.4   GAUSSIAN APPROXIMATION

Another form of approximation is provided by variational methods (Gambling and Matsumura, 1977; Marcuse, 1978; Snyder, 1981; Ankiewicz and Peng, 1992). In this approach, a relatively simple functional dependence is assumed as a trial function for the spatial variation of the fundamental mode. It must also be a reasonable approximation to the exact solution to the problem. The variational technique uses a stationary condition to determine the optimum choice of parameters in the trial function so as to minimize the difference between the approximate and exact solutions.

The fundamental mode field of a weakly guiding waveguide has an approximately parabolic variation within the core, and an exponentially decreasing form in the uniform cladding. This suggests that a Gaussian function is a close approximation to the exact solution over both regions, since it is approximately parabolic close to the waveguide axis. Although it does not have the correct exponential dependence in the cladding, there is only a small fraction of modal power propagating far from the axis, so that for practical V-values immediately below the cutoff of the second mode (see Section 6.3), the error incurred should be minimal. Accordingly, we use a Gaussian transverse dependence as a trial function in an integral form of the scalar wave equation (Snyder and Love, 1983, Chapter 17). By requiring the solution to be stationary, we obtain an equation for the optimal spot size of the Gaussian function. This is the basis of the Gaussian approximation (GA).

4.4.1   Variational equation

The Gaussian approximation for non-circular waveguides expresses the transverse dependence of the fundamental mode field as the product of two Gaussian functions

$$\psi(X, Y) = \exp\left(-\frac{1}{2}\frac{X^2}{S_X^2}\right)\exp\left(-\frac{1}{2}\frac{Y^2}{S_Y^2}\right) \tag{4.21}$$

relative to the normalized coordinates $X = x/\rho_x$, $Y = y/\rho_y$ and normalized spot sizes S_X, S_Y, respectively. To calculate the spot sizes, we use an integral form of the scalar wave equation(Snyder and Love, 1983, Chapter 17):

$$\beta^2 = \frac{\int\int_{A_\infty} \left[k^2 n^2(x,y)\psi^2 - \left(\frac{\partial\psi}{\partial x}\right)^2 - \left(\frac{\partial\psi}{\partial y}\right)^2 \right] dxdy}{\int\int_{A_\infty} \psi^2 dxdy}, \tag{4.22}$$

where A_∞ is the infinite waveguide cross-section. The stationary condition comprises the two equations

$$\frac{\partial\beta^2}{\partial S_X} = 0, \quad \frac{\partial\beta^2}{\partial S_Y} = 0 \tag{4.23}$$

If we substitute (4.21) into (4.22) and rearrange, we obtain

$$U_x U_y = \frac{V_x V_y}{\pi S_X S_Y} \int\int_{S_\infty} f(X,Y) \exp\left[-\left(\frac{X^2}{S_X^2} + \frac{Y^2}{S_Y^2}\right) \right] dXdY +$$
$$\frac{1}{2}\left(\frac{V_y}{V_x}\frac{1}{S_X^2} + \frac{V_x}{V_y}\frac{1}{S_Y^2}\right) \tag{4.24}$$

in terms of the modal parameters U_x and U_y defined by

$$U_x = \rho_x \left(k^2 n_{co}^2 - \beta^2\right)^{1/2}, \quad U_y = \rho_y \left(k^2 n_{co}^2 - \beta^2\right)^{1/2} \tag{4.25}$$

and the normalized index profile $f(X,Y)$, defined by (3.3), and expressed here in the form

$$n^2(X,Y) = n_{co}^2 \left[1 - 2\Delta f(X,Y)\right] \tag{4.26}$$

where $\Delta = (n_{co}^2 - n_{cl}^2)/2n_{co}^2$.

4.4.2 Spot size equations

The profile variation for the step-profile rectangular waveguide is given in terms of a product of Heaviside functions $H(X)$ by

$$f(X,Y) = 1 - [H(X+1) - H(X-1)][H(Y+1) - H(Y-1)] \tag{4.27}$$

along the four boundaries of the core, where $H(Z) = 0$ if $Z < 0$ and $H(Z) = 1$ if $Z > 0$. If we substitute (4.27) into (4.24), we obtain

$$U_x U_y = V_x V_y \left[\text{erfc}(1/S_X) + \text{erfc}(1/S_Y) - \text{erfc}(1/S_X)\text{erfc}(1/S_Y)\right] +$$
$$\frac{1}{2}\left(\frac{V_y}{V_x}\frac{1}{S_X^2} + \frac{V_x}{V_y}\frac{1}{S_Y^2}\right) \tag{4.28}$$

where erfc is the complementary error function defined by

$$\text{erfc}(X) = 1 - \text{erf}(X) = \frac{2}{\sqrt{\pi}} \int_x^\infty e^{-\tau^2} d\tau$$

By expressing β in terms of U_x and U_y in the first and second expressions in (4.23), respectively, the two stationary conditions (4.23) give rise to a pair of coupled transcendental equations for the normalized spot sizes

$$\frac{2V_x^2}{\sqrt{\pi}} \exp\left(-\frac{1}{S_X^2}\right) \mathrm{erf}\left(\frac{1}{S_Y}\right) = \frac{1}{S_X} \tag{4.29a}$$

$$\frac{2V_y^2}{\sqrt{\pi}} \exp\left(-\frac{1}{S_Y^2}\right) \mathrm{erf}\left(\frac{1}{S_X}\right) = \frac{1}{S_Y} \tag{4.29b}$$

For the square-core waveguide, we set $V_x = V_y = V$ and $S_X = S_Y = S$, then both (4.29a) and (4.29b) reduce to

$$\frac{2V^2}{\sqrt{\pi}} \exp\left(-\frac{1}{S^2}\right) \mathrm{erf}\left(\frac{1}{S}\right) = \frac{1}{S} \tag{4.30}$$

The values of S_X, S_Y and S are substituted into (4.28) to determine the corresponding values of U_x, U_y and U, respectively, which in turn determine the value of the propagation constant β.

One of the disadvantages of the Gaussian approximation, like the Marcatili approximation, is the prediction of a spurious cutoff of the fundamental mode of the square guide. As V decreases, the field spreads out in the cladding and the spot size $S \to \infty$. A series expansion in powers of $1/S$ in (4.30) shows that the Gaussian approximation fails below a V-value of

$$V = \frac{\sqrt{\pi}}{2} \approx 0.886$$

However, practical waveguides generally would not be designed with such low V-value.

4.4.3   Accuracy

The Gaussian approximation for the fundamental-mode field of the step-profile rectangular waveguide reduces the solution of the scalar wave equation to the solution of a pair of simple transcendental equations, which further reduce to a single equation for the square-core waveguide. Unlike the Marcatili approximation, which is most accurate as $V \to \infty$, the accuracy of the Gaussian approximation can be expected to be best at intermediate values, as we show in the next chapter. For small values of V, the spread-out, exponentially decreasing field in the cladding is not well approximated by the tail of the Gaussian function, and, likewise, for very large V-values, the confined field in the core approaches the product of the trigonometric functions in (4.3) which differ functionally from the Gaussian product of (4.21).

The range and accuracy of the Gaussian approximation can be improved by using different field representations for the core and cladding regions (Ankiewicz and Peng, 1992) which are closer to the actual field dependence. However, there is the penalty of a loss of simplicity in going to more complicated representations.

REFERENCES

Ankiewicz, A. and Peng, G. D. (1992) Generalized Gaussian approximation for single-mode fibers. *IEEE Journal of Lightwave Technology*, **LT–10**, 22–27.

Buus, J. (1984) Application of the effective index method to nonplanar structures. *IEEE Journal of Quantum Electronics*, **QE–20**, 1106–1109.

Chiang, K. S. (1986) Analysis of optical fibers by the effective-index method. *Applied Optics*, **25**, 348–354.

Gambling, W. A. and Matsumura, H. (1977) Simple characterisation factor for practical single-mode fibres. *Electronics Letters*, **13**, 691–693.

Hocker, G. B. and Burns, W. K. (1977) Mode dispersion in diffused channel waveguides by the effective index method. *Applied Optics*, **16**, 113–118.

Marcatili, E. A. J. (1969) Dielectric rectangular waveguide and directional coupler for integrated optics. *The Bell System Technical Journal*, **48**, 2071–2102.

Marcuse, D. (1978) Gaussian approximation of the fundamental modes of graded-index fibres. *Journal of the Optical Society of America*, **68**, 103–109.

Snyder, A. W. (1981) Understanding monomode optical fibers. *Proceedings of the IEEE*, **69**, 6–13.

Snyder, A. W. and Love, J. D. (1983) *Optical waveguide theory*. London: Chapman & Hall.

5

Numerical methods for modal analysis

As we saw earlier in Chapter 4, there are no exact analytical solutions for the modes of the BCWs being studied, even in the weak-guidance approximation. Accordingly, we must rely on numerical methods to generate accurate solutions of the scalar wave equation for rectangular or square core cross-sections. In this chapter, we review a selection of such methods and describe three of them in detail. For a more comprehensive description of other numerical methods we refer the reader elsewhere (Saad, 1985).

5.1   OVERVIEW OF NUMERICAL METHODS

The first method is based on a decomposition of the modal field into a complete set of sinusoidal basis functions. This method, labelled the *Fourier decomposition method* (FDM) (Henry and Verbeek, 1989), can readily be applied to rectangular- and square-core waveguides, as it is based on a superposition of rectangular domains of constant refractive indices. However, it has the drawback of becoming less and less accurate as a mode approaches its cutoff wavelength.

The second method (Hewlett and Ladouceur, 1995) avoids this limitation by mapping the infinite cross-sectional x–y-plane of the waveguide into a finite domain, and then applying the Fourier decomposition. This method, which leads to accurate values of cutoff wavelength for most waveguiding structures, has been labelled the *modified Fourier decomposition method* (MFDM).

Cutoff wavelengths are fundamental design constraints that have traditionally been difficult to calculate. In order to verify the accuracy of the MFDM we also review, as the third method, the *finite element method* (Yeh *et al.*, 1979; Rahman and Davis, 1984) modified by Chiang (Chiang, 1984; Chiang, 1985; Chiang, 1986) for open (i.e. infinite) domains, and, in Chapter 6, compare its results with those obtained using the MFDM. For completeness, we first refer to other numerical techniques which are used in waveguiding problems.

5.1.1   Point-matching method

One of the oldest numerical methods used to solve the scalar wave equation is the point-matching method. This was used by Goell (Goell, 1969) to solve Maxwell's equations for step-profile square- and rectangular-core waveguides. The basic idea is that, for regions of constant refractive index, there are discrete, separable solutions of the scalar wave equation in cylindrical polar coordinates based on the central axis of the waveguide as the origin in the cross-section. Each such solution is the product of a Bessel function (giving the radial dependence) and a trigonometric function (giving the azimuthal dependence). The general solution in each region consists of the sum over the basic set of these products. For computational purposes, two limited expansions are used: one for the core and one for the cladding. These two expansions and their first derivatives are matched at N points on the core–cladding interface and a system of linear equations is derived for the expansion coefficients and propagation constants; this is solved using standard numerical techniques. The disadvantages of the method are the restriction to step-profile waveguides, and the very large number of coefficients required in the expansions close to mode cutoff.

5.1.2   Harmonic boundary method

The Harmonic boundary method (Skinner, 1985; Ladouceur, 1991) is based on an expansion of the field in terms of a Fourier–Bessel series in the core and cladding, similar to those used by Goell (Goell, 1969). The main difference between these two methods lies in the procedure that transforms these solutions of the scalar wave equation into a set of coupled linear equations. While Goell used point matching of the field on the boundary, the field being evaluated along the boundary is now expanded into a Fourier series and, by matching the core and cladding fields, a set of linear equations for the coefficients of the decomposition is generated. This technique avoids the *ad hoc* choice of the matching points distribution. The method also has the advantage of being able to determine the modal field at cutoff by replacing the Fourier–Bessel expansion over the cladding by cylindrical harmonics, i.e. solutions of the Laplace equation in cylindrical polar coordinates. The major drawback of this method is that, like the point-matching method, it is restricted to step-profile waveguiding structures.

5.1.3   Finite difference methods

The finite difference method is the oldest known method for solving partial differential equations with boundary conditions, and has been used with success in the analysis of non-homogeneous waveguides (Schweig and Bridges, 1984;

Bierwirth *et al.*, 1986; Schulz *et al.*, 1990). Because this method subdivides the domain into many subregions, where the partial differential equation is replaced by a finite difference (or quotient) equation, it lends itself easily to non-homogeneous profiles. The system of linear equations obtained is simpler than those obtained by the finite element method (FEM) (see Section 5.4), and the solution is free from numerical artefacts (the so-called spurious modes) that are observed when using the FEM. The finite difference method is simpler to program than the FEM, and is also simpler to solve, but it offers less flexibility in the modelling of the domain since the subregions used are necessarily square.

5.1.4   Beam propagation method

This method, as its name suggests, solves the scalar wave equation by incremental propagation along the length of a waveguide. It can cope with graded profiles (either absorbing or non-absorbing) and arbitrary cross-sectional geometries. The beam propagation method (BPM) was originally derived for longitudinally-dependent variations, such as tapers and Y-junctions, but is highly inefficient for translationally-invariant waveguides, although all modal quantities of interest can be determined by the method (Feit and Fleck, 1978). This drawback is a consequence of the basic algorithm being based on the full three-dimensional scalar wave equation, rather than the separated two-dimensional equation for the transverse modal dependence. Recently, a variation of the method called *propagation along the imaginary time axis* (Yevick and Hermansson, 1989; Chen and Jüngling, 1994) has made the BPM an efficient tool for modal properties calculation. It can be shown that this approach is equivalent to a class of finite difference methods (Yevick and Bardyszewski, 1992).

Since the original BPM appeared (Feit and Fleck, 1978), numerous other forms have been developed specifically for solving a variety of waveguide and device propagation problems. The original BPM was based on the fast Fourier transform (FFT), whereas more efficient algorithms based on finite difference schemes (Chung and Dagli, 1990) are now generally in use. Further improvements to the finite difference algorithms have been proposed to increase their accuracy (Schmidt, 1993).

Over the period of development of the material and results for this book, different algorithms were used, each of which was the most appropriate at the time. For example, all three-dimensional BPM calculations were carried out using an FFT algorithm, while the bend calculations in Chapter 11 employed a finite difference algorithm, coupled with transparent boundary conditions (Hadley, 1991).

5.2   FOURIER DECOMPOSITION METHOD

Every numerical solution of the scalar wave equation is based on an approximation to the exact solution, together with a criterion which minimizes the error between the two. The various approximate analytical solutions discussed in Chapter 4 also mimimize this error. For example, in the Gaussian approximation of Section 4.4, a stationary expression for the propagation constant is used to determine the spot size of the fundamental mode, while Marcatili's approach relies on the matching of the different approximations to the fields at the interface between the core and the cladding. The Gaussian approximation defines its approximation on the whole x–y-plane, whereas Marcatili's approach divides this plane into domains that support different approximations. The class to which the former approach belongs is called the Galerkin class and the latter the Ritz–Galerkin which includes the finite element method presented in Section 5.4.

Recently, a powerful algorithm, based on a decomposition of the modal fields into a Fourier series, has been proposed (Henry and Verbeek, 1989), and is of particular interest for analyzing square- and rectangular-core waveguides. This method, which we refer to as the Fourier decomposition method (FDM), uses the orthogonality of trigonometric functions to reduce the scalar wave equations to a system of linear equations. An attraction of this algorithm is the subdivision of the domain into rectangular regions where the integration (related to scalar products of trigonometric functions) can be performed analytically.

This method has several advantages over other numerical schemes. Firstly, it is easy to implement because it relies on well-known Fourieranalysis. Secondly, it gives the modal fields of the waveguides in terms on asingle expansion defined on the whole domain which, although itmay seem inconsequential, simplifies the programming task, since each point inthe x–y-plane is treated on an equal footing. Thirdly, this series issimply related to the power in the fields, and, moreover, it offers a common basefor any waveguide profile, hence simplifying the calculation of overlapintegrals between the modal fields of different guides, which is used in later chapters.

5.2.1   Mathematical framework

Our starting point is the scalar wave equation

$$\frac{\partial^2 \psi}{\partial x^2} + \frac{\partial^2 \psi}{\partial y^2} + \left[k^2 n^2(x, y) - \beta^2\right] \psi = 0 \tag{5.1}$$

where $\psi(x, y)$ denotes the transverse modal field, β is the propagation constant and $k = 2\pi/\lambda$, where λ represents the free-space wavelength of the source. In the x–y-plane, we define a rectangular domain with sides of lengths L_x and L_y parallel to the x- and y-axes, respectively, as shown in Figure 5.1. The domain

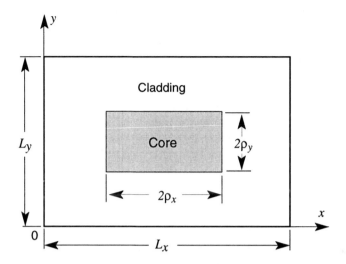

Figure 5.1 Bounding rectangular domain and coordinate system for the Fourier decomposition method.

is assumed to be large enough so that the waveguide core can be located in the centre of the domain with a wide cladding between the core and the nearest boundary, and the modal field can be taken to be zero along the four edges. Next, we assume that the solution of (5.1) within the domain can be expressed as the superposition

$$\psi(x, y) = \sum_{m=1}^{N_m} \sum_{n=1}^{N_n} a_{mn} \phi_{mn}(x, y) \tag{5.2}$$

where the a_{mn} are constants, $N_m, N_n \gg 1$, and the orthonormal basis functions are of the form

$$\phi_{mn}(x, y) = \frac{2}{\sqrt{L_x L_y}} \sin\left(\frac{m\pi x}{L_x}\right) \sin\left(\frac{n\pi y}{L_y}\right) \tag{5.3}$$

where $m = 1, 2, \ldots, N_m$, and $n = 1, 2, \ldots, N_n$.

To simplify the notation and in order to recast the problem into a matrix-eigenvalue problem, we combine the index pair (m, n) into a single index i by means of a suitable one-to-one correspondence. This can be achieved using, for example, the following relations:

$$\begin{aligned} i &= N_n(m - 1) + n \\ m &= (i - 1) \text{ div } N_n + 1 \\ n &= (i - 1) \text{ mod } N_n + 1 \end{aligned} \tag{5.4}$$

where **div** denotes the largest integer not exceeding the ratio $(i-1)/N_n$ and **mod** denotes the difference between $(i-1)/N_n$ and this integer. Since the relation between the pair (m, n) and i is one-to-one, the basis functions obey the orthonormality relation

$$\int_0^{L_x} \int_0^{L_y} \phi_i(x, y)\phi_j(x, y)\, dx dy = \delta_{ij} \tag{5.5}$$

where δ_{ij} is the Kronecker delta function, i.e. $\delta_{ij} = 1$ if $i = j$ and $\delta_{ij} = 0$ otherwise. The scalar wave equation can be transformed into a series of coupled linear algebraic equations by using the orthonormality of the $\phi_i(x, y)$.

5.2.2   Galerkin matrix

Inserting the expansion of (5.2) into (5.1), multiplying by $\phi_j(x, y)$ and integrating over the rectangular domain in the x–y-plane produces the following set of equations

$$\sum_{i=1}^{N_m N_n} \left(V^2 A_{ji} + B_{ji} - W^2 \delta_{ji}\right) a_i = 0 \tag{5.6}$$

where V and W are the waveguide and cladding-mode parameters, respectively

$$V = k\rho \left(n_{co}^2 - n_{cl}^2\right)^{1/2}, \quad W = \rho \left(\beta^2 - k^2 n_{cl}^2\right)^{1/2} \tag{5.7}$$

The length scaling parameter ρ is chosen to be a representative dimension of the waveguide under consideration, n_{co} is normally chosen to be the highest index over the domain and n_{cl} the dominant background index which might differ from the minimum value. The matrix with elements $V^2 A_{ji} + B_{ji}$ is known as the Galerkin matrix, where

$$A_{ji} = \int_0^{L_x} \int_0^{L_y} g(x, y)\phi_j(x, y)\phi_i(x, y)\, dx dy \tag{5.8a}$$

$$B_{ji} = -\left[\left(\frac{m_i \pi \rho}{L_x}\right)^2 + \left(\frac{n_i \pi \rho}{L_y}\right)^2\right]\delta_{ji} \tag{5.8b}$$

and

$$g(x, y) = \frac{n^2(x, y) - n_{cl}^2}{n_{co}^2 - n_{cl}^2} \tag{5.9}$$

is defined by the refractive index profile $n(x, y)$ of the domain. The solution of (5.6) for the a_is determines the scalar transverse field distribution for each mode together with its associated propagation constant β through (5.7). This procedure is equally applicable to waveguides supporting one mode or a finite number of modes. However, as we discussed in the introductory paragraphs of this chapter, the method is inappropriate at and close to cutoff.

5.2.3 General considerations

The linear eigenvalue system of (5.6) is of order $N = N_n N_m$ and is symmetrical. Depending on the geometry of the guide, the sum in (5.2) in the x-direction can be truncated at a value of N_m different from the sum N_n in the y-direction, allowing more flexibility in the method. Similarly, the symmetries of the waveguide can be used to reduce dramatically the size of the eigenvalue system by using only the proper terms in expansion (5.3). The matrix A_{ji} contains off-diagonal elements of the form $\langle \phi_j | g(x,y) | \phi_i \rangle$ that characterize the profile. The explicit form for these elements is

$$\frac{4}{L_x L_y} \int_0^{L_x} \int_0^{L_y} g(x,y) \sin\left(\frac{m_j \pi x}{L_x}\right) \sin\left(\frac{n_j \pi y}{L_y}\right) \times \qquad (5.10)$$

$$\sin\left(\frac{m_i \pi x}{L_x}\right) \sin\left(\frac{n_i \pi y}{L_y}\right) \mathrm{d}x \mathrm{d}y$$

This expression has no explicit simplification for an arbitrary function $g(x,y)$, but in some cases the index profile is composed of regions of uniform index, e.g. the step profile, with consequent simplifications. This suggests an approximation to the index profile by a set of *rectangular regions* of constant refractive index. For such regions, the integral in (5.10) has a closed form solution which can be evaluated analytically by using the identity

$$2 \int_{t_{\min}}^{t_{\max}} \sin(a_i t) \sin(a_j t) \, \mathrm{d}t =$$

$$\begin{cases} t_{\max} - t_{\min} - \dfrac{\sin(2a_i t_{\max}) - \sin(2a_i t_{\min})}{2a_i} & \text{if } a_i = a_j \\[2ex] \dfrac{\sin[(a_i - a_j)t_{\max}] - \sin[(a_i - a_j)t_{\min}]}{a_i - a_j} - & \\[2ex] \dfrac{\sin[(a_i + a_j)t_{\max}] - \sin[(a_i + a_j)t_{\min}]}{a_i + a_j} & \text{if } a_i \neq a_j \end{cases} \qquad (5.11)$$

For guides involving only rectangular or square regions, this process is exact; for different geometries or for graded profiles, one can divide the x–y-plane into many small regions where the uniform index is taken to be the average of the index profile on that region. The integrals in (5.10) depend only on the index profile and not on the V-value, so that these quantities need be calculated only once for a given waveguide. The choice of $g(x,y)$ ensures that in the cladding, where $n = n_{\mathrm{cl}}$, the function $g(x,y) = 0$, and all the integrals in (5.10) vanish. This is particularly convenient for the BCWs and devices considered in this book, since most of the domain in Figure 5.1 consists of the cladding region. Once the subdivision of the domain is made and the matrix elements in (5.8a) are calculated, the system of (5.6) can be determined and solved straightforwardly using any eigensystem library (Smith, 1976).

5.3   MODIFIED FOURIER DECOMPOSITION METHOD

Although the Fourier decomposition method presented in the preceding section is particularly suited for the analysis of rectangular-core or layered waveguides, it suffers from two major drawbacks. Firstly, there exist no concrete rules for determining how far the rectangular bounding box in Figure 5.1 should be placed from the waveguide core (Marcuse, 1991; Gallawa *et al.*, 1992). The dimensions of the box critically affect the accuracy of the results obtained, and usually require optimization by visual inspection of the calculated guided-mode field. Secondly, and perhaps more importantly, the method can only be used to calculate modal field distributions that are reasonably well confined to the core region, i.e. sufficiently far above cutoff. As cutoff is approached, the modal fields spread farther into the cladding and the implicit assumption of zero field at the domain boundary requires a prohibitively large bounding box. Although, in principle, arbitrary accuracy can still be achieved – at least for modes which preserve the zero-field boundary condition at an infinite distance from the core – by increasing the number of spatial frequency components used in the field expansion, the computation time and memory requirements become prohibitively large, before satisfactory convergence is achieved.

To help circumvent these drawbacks, we propose a modified Fourier decomposition method (MFDM) which is accurate down to and including modal cutoff (Hewlett and Ladouceur, 1995). The need for an artificial bounding box is obviated by first mapping the whole of the cartesian x–y plane onto the unit square by means of a suitable transformation. Although this mapping introduces an arbitrary scaling parameter for each transverse dimension, we provide simple physical arguments to show how these parameters can be chosen to achieve optimum convergence rates. Following the development of the preceding section, the modal field is expanded as a complete set of (orthogonal) sinusoidal basis functions in the transformed coordinate system, leading to a standard matrix eigenvalue problem.

5.3.1   Infinite domains

Instead of defining an artificial bounding box as in Figure 5.1, we map the whole of the cartesian x–y-plane onto a unit square in u–v-space via the transformation functions

$$x = \alpha_x \tan\left[\pi\left(u - \frac{1}{2}\right)\right] \tag{5.12a}$$

$$y = \alpha_y \tan\left[\pi\left(v - \frac{1}{2}\right)\right] \tag{5.12b}$$

where α_x, α_y are arbitrary scaling parameters in the x- and y-directions, respectively. This transformation is not unique, but is convenient for our purposes.

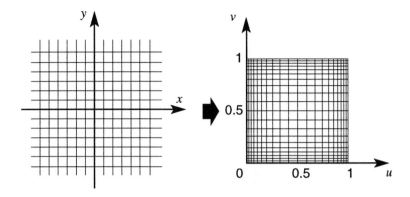

Figure 5.2 Graphical representation of the unit square mapping of(5.12). The infinite x–y-plane is mapped onto the unitsquare in u–v-space.

It has the property that, whilst the core is only slightly compressed in the u–v-plane, the cladding is dramatically compressed. This differential compression is a viable strategy because the modal field only changes slowly at large distances from the core, even close to or at cutoff. The geometrical result of this transformation is illustrated in Figure 5.2. By implementing the same change of variables in the scalar wave (5.1), it becomes

$$\left[\left(\frac{du}{dx}\right)^2\frac{\partial^2}{\partial u^2} + \frac{d^2u}{dx^2}\frac{\partial}{\partial u} + \left(\frac{dv}{dy}\right)^2\frac{\partial^2}{\partial v^2} + \frac{d^2v}{dy^2}\frac{\partial}{\partial v} + \right.$$
$$\left. k^2n^2(u,v) - \beta^2\right]\psi(u,v) = 0 \qquad (5.13)$$

The transverse field component, $\psi(u,v)$, is now expanded as a set of orthogonal sinusoidal basis functions defined over the unit square ofFigure 5.2 to give

$$\psi(u,v) = \sum_{i=1}^{N_m N_n} a_i\phi_i(u,v) = 2\sum_{m=1}^{N_m}\sum_{n=1}^{N_n} a_{mn}\sin(m\pi u)\sin(n\pi v) \qquad (5.14)$$

We collapse the index pair (m,n) into a single index i using the one-to-one correspondence of (5.4) in order to recast the problem into matrix formulation. The Fourier coefficients are denoted a_i, while the integers N_m and N_n represent the total number of Fourier components in the u- and v-directions, respectively, and are chosen to be sufficiently large to ensure adequate convergence.

Equation 5.14 expresses the unknown field, $\psi(u,v)$, as a sinusoidal Fourier series in u–v-space or, equivalently, as a periodic function with odd symmetry about $u = 0$ and $v = 0$ and period 2 in the u- and v-directions. This representation implicitly requires the field to be zero along the boundary of the unit square in Figure 5.2 which, in turn, maps to $x, y = \pm\infty$ under the transformation (5.12). The same mapping also ensures the derivatives $\partial\psi(x, y)/\partial x$ and $\partial\psi(x, y)/\partial y$ vanish on $x = y = \pm\infty$.

5.3.2   Galerkin matrix

Following the derivation of the FDM eigenvalue system presented inSection 5.2, we substitute (5.14) into(5.13), multiply by $\phi_j(u, v)$ and integrate over the unitsquare in Figure 5.2 to obtain

$$\sum_{i=1}^{N_m N_n} \left(V^2 A_{ji} + B_{ji} - W^2 \delta_{ji}\right) a_i = 0 \tag{5.15}$$

which is identical to (5.6), except that the A_{ji} matrix elements are given by (5.8a) and (5.9) with x and $y$replaced by u and v, respectively, and $L_x = L_y = 1$. The B_{ji} matrixelements comprise a non-symmetric contribution to the eigensystem of the form

$$B_{ji} = \rho^2(I_1 + I_2 + I_3 + I_4) \tag{5.16}$$

where the four integrals I_1 to I_4 can be calculated analytically to give

$$I_1 = -m_i^2\pi^2 \int_{u=0}^{1} \int_{v=0}^{1} \left(\frac{du}{dx}\right)^2 \phi_i(u, v)\phi_j(u, v)dudv$$

$$= -\frac{m_i^2}{2\alpha_x^2} \left(\frac{3\delta_{m_i,m_j}}{4} - \frac{\delta_{m_i,m_j-2}}{2} - \frac{\delta_{m_i,m_j+2}}{2} + \frac{\delta_{m_i,2-m_j}}{2} + \right.$$

$$\left. \frac{\delta_{m_i,m_j-4}}{8} + \frac{\delta_{m_i,m_j+4}}{8} - \frac{\delta_{m_i,4-m_j}}{8}\right) \delta_{n_i,n_j} \tag{5.17a}$$

$$I_2 = m_i\pi \int_{u=0}^{1} \int_{v=0}^{1} \left(\frac{d^2u}{dx^2}\right) \frac{1}{\tan(m_i\pi u)}\phi_i(u, v)\phi_j(u, v)dudv$$

$$= \frac{m_i}{\alpha_x^2} \left(\frac{\delta_{m_i,2-m_j}}{4} + \frac{\delta_{m_i,m_j-2}}{4} - \frac{\delta_{m_i,m_j+2}}{4} - \right.$$

$$\left. \frac{\delta_{m_i,4-m_j}}{8} - \frac{\delta_{m_i,m_j-4}}{8} + \frac{\delta_{m_i,m_j+4}}{8}\right) \delta_{n_i,n_j} \tag{5.17b}$$

$$I_3 = -n_i^2\pi^2 \int_{u=0}^{1} \int_{v=0}^{1} \left(\frac{dv}{dy}\right)^2 \phi_i(u, v)\phi_j(u, v) dudv$$

$$= -\frac{n_i^2}{2\alpha_y^2} \left(\frac{3\delta_{n_i,n_j}}{4} - \frac{\delta_{n_i,n_j-2}}{2} - \frac{\delta_{n_i,n_j+2}}{2} + \frac{\delta_{n_i,2-n_j}}{2} + \right.$$

$$\left. \frac{\delta_{n_i,n_j-4}}{8} + \frac{\delta_{n_i,n_j+4}}{8} - \frac{\delta_{n_i,4-n_j}}{8} \right) \delta_{m_i,m_j} \tag{5.17c}$$

$$I_4 = n_i\pi \int_{u=0}^{1} \int_{v=0}^{1} \left(\frac{d^2v}{dy^2} \right) \frac{1}{\tan(n_i\pi v)} \phi_i(u,v)\phi_j(u,v)\,du\,dv$$

$$= \frac{n_i}{\alpha_y^2} \left(\frac{\delta_{n_i,2-n_j}}{4} + \frac{\delta_{n_i,n_j-2}}{4} - \frac{\delta_{n_i,n_j+2}}{4} - \right.$$

$$\left. \frac{\delta_{n_i,4-n_j}}{8} - \frac{\delta_{n_i,n_j-4}}{8} + \frac{\delta_{n_i,n_j+4}}{8} \right) \delta_{m_i,m_j} \tag{5.17d}$$

where δ_{ij} is the Kronecker delta function.

Equation 5.15 represents a standard matrix eigenvalue problem which can be readily solved using standard numerical library routines (Smith, 1976). In particular, the eigenvalues, W^2, and eigenvectors, $\mathbf{a} = (a_1, a_2, \dots \ a_{N_m N_n})^T$, of the matrix $V^2 A_{ji} + B_{ji}$ determine the modal propagation constants and the associated Fourier expansion coefficients, respectively. The modal fields can then be readily represented in u–v-space via (5.14) and mapped back to the x–y-space with the transformation functions (5.12).

Two helpful properties should be noted when calculating the matrix elements. Firstly, rectangular regions of constant refractive index in x–y-space transform to rectangular regions of the same refractive index in u–v-space. For such regions, the integral in (5.8a) can then be evaluated analytically by using (5.11). In contrast, non-rectangular regions of constant refractive index in x–y-space become distorted under the transformation (5.12) and analytical expressions are not generally available for (5.8a). Numerical integration can, however, be avoided by approximating such regions by a set of thin rectangular elements and iteratively applying (5.11). Arbitrary core geometries can easily be incorporated using this technique.

Secondly, the order of the system matrix with elements $V^2 A_{ji} + B_{ji}$ can be considerably reduced by taking advantage of modal symmetry if it is known *a priori*. For example, a mode with even symmetry about $x = 0$ in the x–y-plane will contain only Fourier components which possess even symmetry about $u = 1/2$ in transformed coordinates. Conversely, a mode with odd symmetry about $x = 0$ in the x–y-plane will contain only Fourier components which possess odd symmetry about $u = 1/2$ in transformed coordinates. Analogous results hold for modal symmetries about $y = 0$. Finally, we note that the system matrix elements $V^2 A_{j,i} + B_{j,i}$ are real and asymmetric, and may therefore possess complex eigenvalues and eigenvectors. The *bound* modes of interest will, however, correspond to the real, positive eigenvalues and their associated eigenvectors (Snyder and Love, 1983, Section 11–17).

5.3.3   Modal cutoff

Given the cross-sectional geometry, profile and associated V-value of a wave-guide, the cladding mode parameter, W, of all bound modes can be readily calculated using the method discussed above, subject to the implicit boundary condition that the field remains zero at infinity. We now show how to calculate the cutoff V-value, V_{co}, of each mode directly.

At modal cutoff the propagation constant $\beta = kn_{cl}$ (Snyder and Love, 1983, Section 11–18). Thus, $W = 0$ from (5.7) and (5.17) reduces to the generalized eigenvalue problem:

$$\sum_{i=1}^{N_m N_n} \left(V_{co}^2 A_{ji} + B_{ji} \right) a_i = 0 \qquad (5.18)$$

which is readily solved using standard numerical library routines. In particular, the eigenvalues, $1/V_{co}^2$, and eigenvectors, $\mathbf{a} = (a_1, a_2, \ldots, a_{N_m N_n})^T$, of the system matrix $-B_{ji}^{-1} A_{ji}$ determine the cutoff V-values and the associated Fourier expansion coefficients, respectively. It is then straightforward to reconstruct the modal fields at cutoff via (5.14) and the transformation functions (5.12).

5.3.4   Choice of scaling parameters

The transformation equations (5.12) involve two arbitrary scaling parameters, α_x and α_y, corresponding to different mapping compression of the x- and y-directions. If these parameters are not chosen judiciously, an unnecessarily large number of Fourier components may be required in the field expansion of (5.14) to achieve satisfactory convergence. To help minimize the number of Fourier components and ensure near-optimum convergence rates, α_x and α_y should be chosen such that the rapidly varying parts of the modal field are evenly distributed across the transformed coordinate system. As with all Fourier expansions, steep slopes imply that many terms are needed for a given accuracy. Accordingly, we try to minimize the maximum slope of $\phi(x, y)$ in the transformed coordinate system. This procedure can be illustrated by the following example.

Consider the waveguide depicted in Figure 5.1 with a square-core, i.e. $\rho_x = \rho_y = \rho$. If the V-value is large enough for the waveguide to be two-moded, this waveguide can support a mode of odd symmetry in the x-direction and even symmetry in the y-direction, usually referred to as the LP_{11}^{oe} mode. At cutoff, the two extrema in the field distribution of this mode lie close to the core–cladding interface at $x = \pm\rho$. This property is illustrated in Figure 5.3, which shows, for various values of α_x, cuts along the line $v = 1/2$ of the modal field profile at cutoff in the transformed coordinate system. In each case, the peaks in the field distribution lie close to the pairs of vertical lines which indicate the

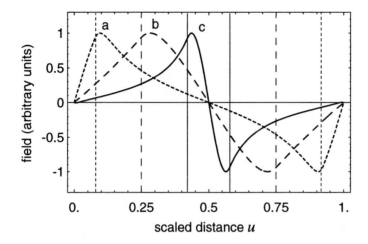

Figure 5.3 Field distributions of the LP$_{11}^{oe}$ mode at cutoff for (a) $\alpha_x = \rho_x/4$, (b) $\alpha_x = \rho_x$ and (c) $\alpha_x = 4\rho_x$. Each plot shows the field profile along $v = 1/2$ in the transformed coordinate system. The position of the core–cladding interface is indicated by the pairs of vertical solid lines. The simulation parameters correspond to $n_{co} = 1.45$, $n_{cl} = 1.447$ and $\rho_x = \rho_y = 4$ μm, with $\rho = \sqrt{\rho_x \rho_y}$.

position of the core–cladding interface. The simulation parameter values are given in the figure caption.

It is clear from these plots that when $\alpha_x < \rho_x$ (curve a), the field profile varies most rapidly in the cladding region, while for $\alpha_x > \rho_x$ (curve c), the rapid variation occurs within the waveguide core. In the case $\alpha_x = \rho_x$ (curve b), the core–cladding interfaces map to $u = 1/4$ and $u = 3/4$, and the field peaks and zeros are almost regularly spaced in the transformed coordinate system and there are no regions of steep profile. This situation provides the most gradual variation across the entire field profile and essentially minimizes the number of Fourier components required in the field expansion. Near-optimum convergence rates are then achieved for this particular mode. More generally, rapid convergence is obtained by setting $\alpha_x = \rho_x$ and $\alpha_y = \rho_y$ in most situations of practical interest.

5.4   FINITE ELEMENT METHOD

The finite element method (FEM) has been very successful in the numerical solution of partial differential equations because of its ability to model the most intricate domain geometries (Schwarz, 1988). It extends the concept of the finite difference method subdivision of the domain to arbitrary shapes (elements) where basic functions are defined according to the shape of the element. The versatility of the method, however, requires a relatively greater complexity in programming and bigger demands on both computer time and memory. Another advantage of the FEM is that it can determine modal parameters accurately close to cutoff (Chiang, 1984; Chiang, 1985; Silvester et al., 1977). The problem of spurious modes (Rahman and Davis, 1984; Hayata et al., 1986) need not concern us, since we are predominantly interested in the BCW monomode régime.

One drawback of the FEM is its relative programming complexity and the unwieldy field representation it generates. Different fields calculated on different meshes are difficult to overlap, and quantities such as energy and energy densities are cumbersome to calculate. We believe that the MFDM of Section 5.3 is a more versatile calculation tool for the waveguiding structures produced by an etching/deposition process. Therefore, the FEM will be used primarily for the determination of cutoff in order to make useful numerical comparisons with the corresponding results of the MFDM. The following sections give only an outline of the FEM.

5.4.1   Variational approach

The finite element method is based on the minimization of a given functional related to the scalar wave equation (Chiang, 1985). If we express this equation in the form

$$\left\{ \frac{\partial^2}{\partial x^2} + \frac{\partial^2}{\partial y^2} + k^2 \left[n^2(x,y) - n_{cl}^2 \right] - w^2 \right\} \psi(x,y) = 0 \qquad (5.19)$$

where w is the un-normalized cladding mode parameter defined as

$$w = \left(\beta^2 - k^2 n_{cl}^2 \right)^{1/2} \qquad (5.20)$$

then the functional is

$$I = \frac{1}{2} \int \int_A \left\{ \left(\frac{\partial \psi}{\partial x} \right)^2 + \left(\frac{\partial \psi}{\partial y} \right)^2 - \left[k^2 \left(n^2(x,y) - n_{cl}^2 \right) + w^2 \right] \psi^2 \right\} dxdy \qquad (5.21)$$

which is stationary under the associated condition $\psi = \psi(s)$ on the boundary of the domain of A in the waveguide cross-section, where s describes the boundary path. Working with (5.19) and approximating the partial derivatives by finite differences leads to the class of methods known as finite difference methods. Approximating (5.21) by numerical quadrature formula, e.g. Simpson's rule, and subdividing the domain into elements over which the formula can be calculated analytically, leads to the finite element method.

5.4.2   Discretization

The domain A in the waveguide cross-section is first discretized. The shape of the resulting elements can be chosen quite freely to follow the boundaries of the domain under study. The most commonly-chosen elements are triangular, but are in no way restricted to this shape. Even elements with curvilinear sides and multiple vertices are permitted. The field solution of (5.19) is approximated on each element by a basis polynomial $\alpha(x, y)$ with coefficients depending linearly on the values of the field $\psi(x_i, y_i) = \psi_i$, where (x_i, y_i) are the coordinates of node i. Second-order basis polynomials defined over triangular elements are often used, as they both improve the accuracy and reduce the number of elements for a given domain without introducing too many computational complications. Once the basis polynomials are chosen, (5.21) can be integrated analytically on each element to obtain a series of linear relations between the coefficients of the basis polynomials defined on each element.

Once assembled, these relations give a quadratic expression in terms of the value of the field ψ_i for the stationary integral (5.21). The stationary conditions then imply that $\partial I/\partial \psi_i = 0$, hence producing a set of linear equations. The order of the matrices involved is equal, in this case, to the number of nodes introduced in the triangulation of the domain. The FEM matrix equation equivalent to (5.21) is given by

$$P\psi - k^2 Q\psi = 0 \tag{5.22a}$$

where

$$P = S + w^2 T \tag{5.22b}$$

$$S_{ij} = \sum_e \int_{A_e} \nabla \alpha_i \nabla \alpha_j \, dx dy \tag{5.22c}$$

$$T_{ij} = \sum_e \int_{A_e} \alpha_i \alpha_j \, dx dy \tag{5.22d}$$

$$Q_{ij} = \sum_e \int_{A_e} [n^2(x, y) - n_{cl}^2] \alpha_i \alpha_j \, dx dy \tag{5.22e}$$

The summation $\sum_e$ runs over all elements, and A_e is the area of each element. An example of triangulation of a square domain is given in Figure 5.4. The system (5.22) is known as a generalized eigenvalue problem, and can be solved by first reducing it to the canonical form $P'\psi = k^2\psi$ using a Cholesky decomposition (Stoer and Bulirsch, 1980). The complexity in using the FEM is directly proportional to the size of the domain and the resolution of the discretization grid.

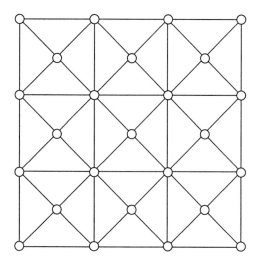

Figure 5.4 Triangulation of a square domain. There are 25 nodes belonging to 36 elements, forming a linear system of equations of order 25. If homogeneous boundary conditions are applied, the order of the system is reduced to 13, the number of inner nodes.

5.4.3   Infinite domains

In order to solve the scalar wave equation for a mode near its cutoff value, the size of the grid must be chosen to be very large, due to the spreading of the field, in order for the boundary conditions to be realistically applied. This difficulty has been overcome by generalizing the FEM to infinite domains (Silvester *et al.*, 1977), and has been applied to waveguiding structures (Chiang, 1985). The main idea is to map the initial boundary C_0 of the domain onto a new boundary C_i, exterior to the initial domain, by the mapping

$$C_i \leftarrow K^i C_0 \qquad (5.23)$$

This process is illustrated in Figure 5.5.

The grey annular region in Figure 5.5 is then discretized and the system (5.22) is constructed for that region. A second annular region, C_2, is constructed, discretized and the system in (5.22) modified to take into account this second annular region. The next step is to eliminate the nodes on C_1 in Figure 5.5 numbered from (6) to (10), and to condense their effect on to the nodes of C_0 and C_2. This condensation process is iterated until the boundary C_i encloses a domain large enough for the boundary condition to be applied. For example, if the mapping constant $K = 1.2$ and 40 condensations are applied, a point on C_0

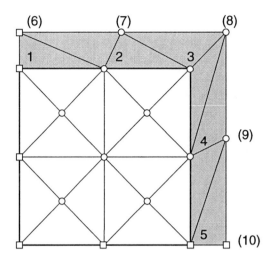

Figure 5.5 Construction of the first ring (grey area) around the upper right quarter of a square domain (e.g. the core of a square guide). The mapping constant $K = 1.2$, and the nodes represented by squares are the loci of boundary conditions appropriate for the particular mode.

at unit distance from the origin would be projected at a distance $K^{40} \approx 1470$. This iteration scheme clearly has an advantage over brute discretization of a domain of this size.

REFERENCES

Bierwirth, K., Schulz, N. and Arndt, F. (1986) Finite-difference analysis of rectangular dielectric waveguide structures. *IEEE Transactions on Microwave Theory and Techniques*, **MTT–34**, 1104–1114.

Chen, J. C. and Jüngling, S. (1994) Computation of higher-order waveguide modes by imaginary-distance beam propagation method. *Optical and Quantum Electronics*, **26**, S199–S205.

Chiang, K. S. (1984) Finite element method for cutoff frequencies of weakly guiding fibres of arbitrary cross-section. *Optical and Quantum Electronics*, **16**, 487–493.

Chiang, K. S. (1985) Finite-element analysis of optical fibres with iterative treatment of the infinite 2-D space. *Optical and Quantum Electronics*, **17**, 381–391.

Chiang, K. S. (1986) Finite element analysis of weakly guiding fibers with arbitrary refractive-index distribution. *IEEE Journal of Lightwave Technology*, **LT–4**, 980–990.

Chung, Y. and Dagli, N. (1990) An assessment of Finite Difference Beam Propagation Method. *IEEE Journal of Quantum Electronics*, **QE–26**, 1335–1339.

Feit, M. D. and Fleck, J. A. Jr. (1978) Light propagation in graded-index optical fibers. *Applied Optics*, **17**, 3990–3998.

Gallawa, R. L., Goyal, I. C. and Ghatak, A. K. (1992) Modal properties of circular and noncircular optical waveguides. *Fiber and Integrated Optics*, **11**, 25–50.

Goell, J. E. (1969) A circular-harmonic computer analysis of rectangular dielectric waveguides. *The Bell System Technical Journal*, **48**, 2133–2160.

Hadley, G. R. (1991) Transparent boundary condition for beam propagation. *Optics Letters*, **16**, 624–626.

Hayata, K., Koshiba, M., Egushi, M. and Suzuki, M. (1986) Novel finite-element formulation without any spurious solutions for dielectric waveguides. *Electronics Letters*, **22**, 295–296.

Henry, C. H. and Verbeek, B. H. (1989) Solution of the scalar wave equation for arbitrarily shaped dielectric waveguides by two-dimensional Fourier analysis. *IEEE Journal of Lightwave Technology*, **LT–7**, 308–313.

Hewlett, S. J. and Ladouceur, F. (1995) Fourier Decomposition Method applied to mapped infinite domains: scalar analysis of dielectric waveguides down to modal cut-off. *IEEE Journal of Lightwave Technology*, **LT–13**, 375–383.

Ladouceur, F. (1991). *Buried channel waveguides and devices*. Ph.D. thesis, Australian National University.

Marcuse, D. (1991) *Theory of dielectric optical waveguides, 2nd edition*. San Diego: Academic Press.

Rahman, B. M. A. and Davis, J. B. (1984) Finite-element solution of integrated optical waveguides. *IEEE Journal of Lightwave Technology*, **LT–2**, 682–688.

Saad, S. M. (1985) Review of numerical methods for the analysis of arbitrarily-shaped microwave and optical dielectric waveguides. *IEEE Transactions on Microwave Theory and Techniques*, **MTT–33**, 894–899.

Schmidt, F. (1993) An adaptative approach to the numerical solution of Fresnel's wave equation. *IEEE Journal of Lightwave Technology*, **LT–11**, 1425–1434.

Schulz, N., Bierwirth, K., Arndt, F. and Köster, U. (1990) Finite-difference method without spurious solutions for the hybrid-mode analysis of diffused channel waveguides. *IEEE Transactions on Microwave Theory and Techniques*, **MTT–38**, 722–729.

Schwarz, H. R. (1988) *Finite element methods*. London: Academic Press.

Schweig, E. and Bridges, W. B. (1984) Computer analysis of dielectric waveguides: a finite-difference method. *IEEE Transactions on Microwave Theory and Techniques*, **MTT–32**, 531–541.

Silvester, P. P., Lowther, D. A., Carpenter, C. J. and Wyatt, E. A. (1977) Exterior finite elements for 2-dimensional field problems with open boundaries. *Proceedings IEE*, **124**, 1267–1270.

Skinner, I. M. (1985). *Studies of axi-symmetric dielectric waveguides*. Ph.D. thesis, Australian National University.

Smith, B. T. (1976) *Matrix eigensystem routines – EISPACK guide*. New York: Springer.

Snyder, A. W. and Love, J. D. (1983) *Optical waveguide theory*. London: Chapman & Hall.

Stoer, J. and Bulirsch, R. (1980) *Introduction to numerical analysis*. New York: Springer-Verlag.

Yeh, C., Ha, K., Dong, S. B. and Brown, W. P. (1979) Single-mode optical waveguides. *Applied Optics*, **18**, 1490–1504.

Yevick, D. and Bardyszewski, W. (1992) Correspondence of variational finite-difference (relaxation) and imaginary-distance propagation methods for modal analysis. *Optics Letters*, **17**, 329–330.

Yevick, D. and Hermansson, B. (1989) New formulation of the matrix beam propagation method: application to rib waveguides. *IEEE Journal of Quantum Electronics*, **QE–25**, 221–229.

6
Modes of buried channel waveguides

In this chapter, we bring together results derived from the approximation methods of Chapter 3 and from the numerical methods of Chapter 5. For convenience, we concentrate on the fundamental modes of the step-profile square- and rectangular-core BCWs, although all the methods are adaptable to graded profiles of arbitrary cross-sectional geometry. The emphasis is on the modal propagation constant β, expressed in normalized form in terms of the core modal parameter U through

$$U = \rho \left(k^2 n_{\text{co}}^2 - \beta^2 \right)^{1/2} \tag{6.1}$$

and the corresponding plane-polarized electric field. For the square-core BCW, the normalized eigenvalue U depends solely on the value of V, where

$$V = k\rho (n_{\text{co}}^2 - n_{\text{cl}}^2)^{1/2} \tag{6.2}$$

and $k = 2\pi/\lambda$, so that a single set of U- vs V-values for the fundamental mode presents a universal result valid for any combination of source wavelength λ, core side 2ρ, and core and cladding refractive indices n_{co} and n_{cl}, respectively. The corresponding fundamental-mode results for the rectangular-core BCW depend on both V and the core aspect ratio. Section 6.3 determines the second-mode cutoff for these BCWs, which defines the single-mode régime that must be maintained for compatibility with single-mode telecommunications fibre.

6.1 APPROXIMATION METHODS

6.1.1 Marcatili approximation

Within the Marcatili approximation of Section 4.2, the U-value is given explicitly in terms of the corresponding slab waveguide value of U_{sl} by (4.14). The latter is determined numerically in terms of V through (4.12) with $W_x = W$ and $V_x = V$. However, since this approximation is most accurate for larger values of V, it is sufficient to use the explicit solution of (4.17). Results are plotted in Figure 6.1. Note that there is a spurious cutoff when $U = V = V_{\text{co}}$, giving $V_{\text{co}} = \pi/\sqrt{2} - 1 \approx 1.221$ from (4.17). Apart from its inherent inaccuracy for smaller values of V, the Marcatili approximation is not applicable

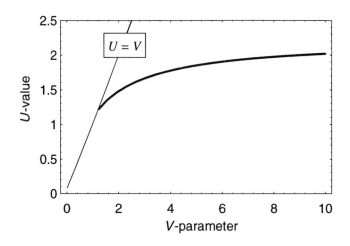

Figure 6.1 U as a function of V for the fundamental mode of a square-core BCW based on the Marcatili approximation. The line $U = V$ denotes cutoff.

to graded-profile BCWs, although it can be readily extended to rectangular BCWs.

6.1.2   Effective index method

The effective index method is the only approximation method presented here that does not produce a pseudo cutoff. For its application to square-core BCWs, the core parameter U is calculated in two stages. First, one of (4.20) is solved, say for the x-direction, giving an intermediate eigenvalue U_x. This value is then used to produce an *effective index* $n_{\text{eff}} = \beta/k$ through the relation

$$U_x = k\rho \left(n_{\text{co}}^2 - n_{\text{eff}}^2\right)^{1/2}$$

This value of n_{eff} replaces the core value when solving for the y-direction, hence producing an effective V-value V_{eff}. Finally, (4.20) is solved in the y-direction using V_{eff}, giving an effective U_y-value. The calculated eigenvalues U_x and U_y are then substituted into (4.10), where V_y is replaced by V_{eff}. The numerical solution is illustrated in Figure 6.2.

In addition to the simple example given above, the effective index method is also used for a variety of guiding structures in integrated optics created by etching and deposition processes. It is applicable to a wide class of problems, including layered guiding structures, and is a powerful tool for reducing a two-dimensional problem to two coupled one-dimensional problems, leading

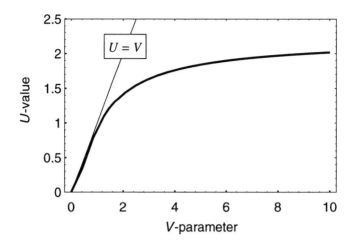

Figure 6.2 U as a function of V for the fundamental mode of a square-core BCW based on the effective index method. The line $U = V$ denotes cutoff.

to a simplification in programming and reduced computer processing time. Although the eigenvalues calculated with this method are accurate, there is a major drawback in calculating the field. Since the calculation involves different core-indices in the x- and y-directions, the field determined by this method is asymmetric, whereas if the waveguide is square, the field is symmetric.

6.1.3   Gaussian approximation

The value of U for the fundamental mode within the Gaussian approximation (GA) is determined in two stages. Given the value of V, we first solve (4.30) for the normalized spot size, S, numerically using, for example, the Newton–Raphson method (Press *et al.*, 1988). Both S and V are then substituted into (4.28) with $S = S_x = S_y$ and $V = V_x = V_y$, to determine U explicitly:

$$U^2 = V^2 \left[1 - \text{erf}^2(1/S)\right] + 1/S^2 \tag{6.3}$$

A plot of U against V is given in Figure 6.3 and a table of U- and V-values is provided in Table 6.1. Like the Marcatili approximation, the GA predicts a spurious cutoff, this time at $U = V_{\text{co}} = \sqrt{\pi}/2 \approx 0.886$. This arises because, as V decreases, the Gaussian function, with its quadratic exponent, becomes a poorer approximation to the actual field in the cladding with its linear exponent. The generalized Gaussian approximation (Ankiewicz and Peng, 1992)

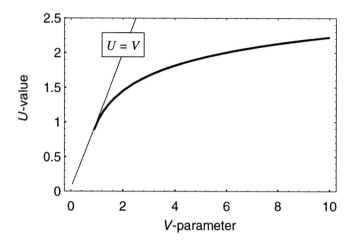

Figure 6.3 U as a function of V for the fundamental mode of a square-core BCW based on the GA. The line $U = V$ denotes cutoff.

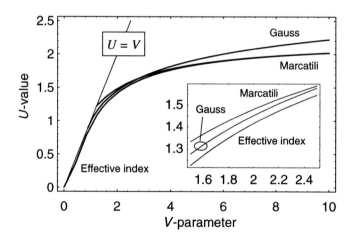

Figure 6.4 U-curves for the fundamental mode of a square-core BCW comparing the various approximations.

overcomes this problem. For comparison, the Marcatili, Gaussian and effective index approximation curves are plotted together in Figure 6.4.

6.2   NUMERICAL METHODS

6.2.1   Modified Fourier decomposition method

The MFDM of Section 5.3 is applied to the square-core waveguide with $N_m = N_n$ in (5.14). The scaling coefficients α_x and α_y of the mapping (5.12) are chosen – as discussed in Section 5.3.4 – to be equal to the core half-width ρ, hence ensuring optimal convergence. The symmetry of the structure (and of the fundamental mode) was taken into account and we have chosen the number of even waves, N_m and N_n, sufficiently large to ensure convergence of the calculated normalized modal parameter $U(V)$. Table 6.1 lists these values, and, for comparison, the corresponding U-values calculated by the FEM and GA are included.

There are two limiting cases which strongly affect the accuracy of this calculation. For large values of V, the fundamental-mode field is confined predominantly within the core of the waveguide. In the limit $V \to \infty$, the field is totally confined within the core and a discontinuity appears in the slope of the field at the core–cladding interface. This abrupt variation for large V-values is represented by a spectrum of high spatial frequencies in the Fourier decomposition and hence requires $N_m, N_n \gg 1$, leading to long calculation times and large computer memory requirements.

At the opposite extreme, for sufficiently small V-values, the field spreads out farther into the cladding but, at least for the fundamental and higher-order modes with the same symmetry, the appropriate boundary condition to apply at infinity on the field is zero first derivative, rather than zero field. The MFDM (Hewlett and Ladouceur, 1995) can be readily adapted to include the limit $V \to 0$ without any loss of accuracy.

The comparison of the MFDM, FEM and GA results of Table 6.1 for U are presented in the curves of Figure 6.5. These show the increasing discrepancy between the GA and the more accurate MFDM values as V increases. This is attributable to the difference between the product of the two Gaussian functions and the product of the two trigonometric functions for the exact solution of the (4.3) in the perfectly conducting limit. The MFDM and FEM solutions are in close agreement and the slight discrepancy between them can be reduced by increasing the number of elements and calculation time.

Figure 6.6 shows the contour lines of the fundamental mode field of the square-core step-index BCW for the parameters values indicated in the caption. The figure shows the almost circular symmetry of the field within the core imposed by the four-fold symmetry of the refractive index profile. The field in the cladding loses its circular symmetry farther away from the core–cladding interface where the field amplitude decreases rapidly.

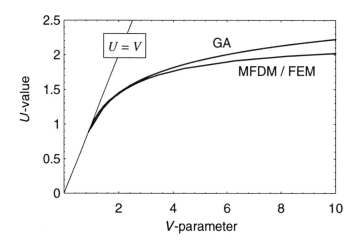

Figure 6.5 Comparison of finite element (FEM), modified Fourier decomposition (MFDM), and Gaussian approximation (GA) methods for the step-profile, square-core BCW.

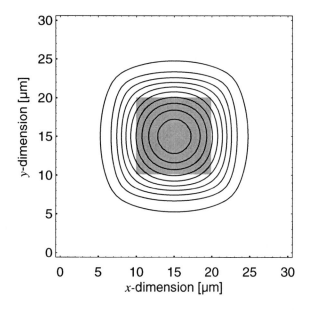

Figure 6.6 Contours lines of the fundamental mode field of the square-core step-index BCW of core half-width $\rho = 5.0$, core index $n_{co} = 1.450$, cladding index $n_{cl} = 1.447$ and wavelength $\lambda = 1.3 \ \mu m$.

Table 6.1 Comparison of modified Fourier decomposition (MFDM), finite element (FEM) and Gaussian approximation (GA) methods for the case of the step-profile, square-core BCW

V	U: MFDM	U: FEM	U: Gauss
10.0	2.019482	2.02025	2.22104
9.00	1.998884	1.99912	2.17762
8.00	1.973686	1.97462	2.12822
7.00	1.942291	1.94319	2.07105
6.00	1.901456	1.90234	2.00341
5.00	1.846596	1.84787	1.92092
4.00	1.768723	1.76924	1.81589
3.00	1.648814	1.64887	1.67274
2.50	1.560325	1.56033	1.57666
2.40	1.539082	1.53910	1.55449
2.30	1.516371	1.51638	1.53110
2.20	1.492028	1.49205	1.50635
2.10	1.465866	1.46588	1.48009
2.00	1.437668	1.43770	1.45213
1.90	1.407181	1.40720	1.42225
1.80	1.374109	1.37415	1.39018
1.70	1.338105	1.33814	1.27715
1.60	1.298763	1.29883	1.23212
1.50	1.255603	1.25568	1.18214
1.00	0.960108	0.95993	–

6.2.2 Finite element method

Although the MFDM is a most convenient method for rectangular- and square-core waveguides, it is useful to make a comparison with the FEM in order to compare the two methods. Accordingly, we present in Table 6.1 values of U calculated by the two methods for the same V-value. As explained in Section 5.4, these calculations were performed using a discretization of the upper left quadrant of the core with 64 nodes and 77 elements, and an iteration scheme with mapping constant $K = 1.2$ and 40 condensations to model the unbound cladding. If the MFDM results are regarded as essentially exact, the maximum relative error in the U-values is 0.02% over the V-value range of Table 6.1. This difference can be decreased at the expense of increasing the number of elements in the FEM and increasing the number of base functions in the MFDM.

6.3   SECOND-MODE CUTOFF

The range of wavelengths for single-mode operation of rectangular and square-core BCWs needs to be established in order for these waveguides to be compatible with single-mode telecommunication fibres. There is no exact analytical solution to this problem, but a technique has been devised to give a useful estimate of the cutoff wavelength; this is discussed below. Furthermore, there are few numerical methods which are sufficiently accurate to determine cutoff wavelength for arbitrary profiles. These include the MFDM and FEM methods, which are applied below to the second-mode cutoff of square and rectangular-core, step-profile BCWs.

6.3.1   Approximation method

A BCW possesses two second-order modes, which can be denoted by LP_{11}^{eo} and LP_{11}^{oe}, and are illustrated schematically in Figure 6.7 for a rectangular-core BCW. The superscript refers to the symmetry of the modes relative to the x- and y-axes, respectively, while the first subscript refers to the azimuthal anti-symmetry of the mode and the second subscript refers to the lowest-order mode, i.e. the largest value of propagation constant, with this symmetry. The single-mode region of operation for the BCW is characterized by the cutoff V-value of the LP_{11}^{oe} mode, whose propagation constant value is closest to that of the fundamental mode. For BCWs with arbitrary-shaped profiles and cross-sections, the LP_{11}^{eo} and LP_{11}^{oe} modes generally have distinct propagation constants, but they are degenerate for both circular and square cross-sections, and, therefore, have the same cutoff V-value, V_{co}.

A knowledge of the LP_{11} mode cutoff value V_{co}^{f} for the circular fibre can be used to generate an accurate approximation to the corresponding cutoff value V_{co}^{s} for the square-core waveguide (Sammut, 1982). The derivation is based on the perturbation formula for the scalar wave equation (Snyder and Love, 1983, Equation 18-4)

$$\beta_f^2 - \beta_s^2 = k^2 \int\int_{-\infty}^{\infty} (n_f^2 - n_s^2)\psi_f\psi_s \mathrm{d}x\mathrm{d}y / \int\int_{-\infty}^{\infty} \psi_f\psi_s \mathrm{d}x\mathrm{d}y \qquad (6.4)$$

where β is the propagation constant, n the refractive-index profile, ψ the solution of the scalar wave equation, and subscripts f and s refer to the fibre and square-core waveguide, respectively. This relationship also provides a way to define an equivalent circular-core fibre to the square-core waveguide so that corresponding modes have the same propagation constant. Thus we set $\beta_f = \beta_s$ in (6.4) whence

$$0 = \int\int_{-\infty}^{\infty} (n_f^2 - n_s^2)\psi_f\psi_s \mathrm{d}x\mathrm{d}y \qquad (6.5)$$

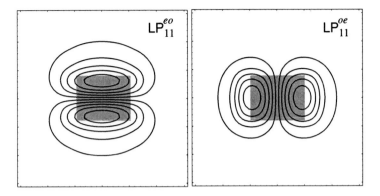

Figure 6.7 Schematic of field contour lines of the LP_{11} modes for the rectangular-core BCW.

If the fibre and waveguide have the same profiles, e.g. the step profile, then $n_f = n_s$ everywhere except in the shaded regions of Figure 6.8 where the cores overlap. If ψ_f and ψ_s denote the second-mode fields of the fibre and BCW, respectively, then we assume these fields have the same radial dependence, and a common sinusoidal azimuthal dependence. This is a reasonable approximation, considering the similar symmetries of the square- and circular-core cross-section. The contributions to the integral in (6.5) from the four regions inside the circle will then cancel the four contributions from the regions outside the circle provided the areas of the circular and square cores are equal. This applies, in particular, at the cutoff of the second mode (Sammut, 1982). Thus a first approximation is given by

$$U_s^{co} = V_s^{co} \simeq \frac{\sqrt{\pi}}{2} U_f^{co} = 2.131 \tag{6.6}$$

The equal-area result for rectangular cores becomes increasingly inaccurate as the aspect ratio (ratio of longer to shorter sides) increases. This arises because the assumption of constant values of the scalar wave equation solutions in the overlap areas between the circular and rectangular interfaces is no longer valid for long thin rectangles which extend far into the cladding of the circular fibre.

6.3.2   Numerical methods

As discussed in the previous chapters, both the MFDM and FEM give accurate values for U both close to and at cutoff. The second-mode cutoff is found by imposing the zero boundary condition at infinity, for the MFDM, and sufficiently

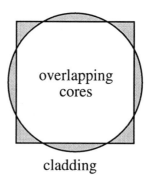

Figure 6.8 Superposed cross-sections of fibre and square-core BCW index profiles. The shaded regions indicate where the two profiles differ.

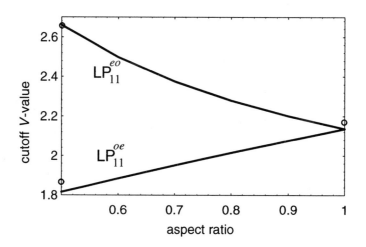

Figure 6.9 Cutoff V-values for the LP^o_{11} and LP^e_{11} modes of a step-profile rectangular-core waveguide as a function of the aspect ratio $\epsilon = \rho_y/\rho_x$.

far from the core in the case of the FEM. The cutoff calculations use here the same values of parameters as those used in Section 6.2.1.

Figure 6.9 shows the cutoff V-values for the LP^{eo}_{11} and LP^{oe}_{11} modes of a step-profile rectangular-core BCW as a function of the aspect ratio $\epsilon = \rho_y/\rho_x$, calculated by both the MFDM and FEM. The good agreement between the two methods can be deduced from Table 6.2 and is consistently within 0.1% for aspect ratio ϵ in the range 0.5 to 2.0. Note that $V_{co} = 2.13$ for the square core, in excellent agreement with the approximate value given by (6.6).

Table 6.2 LP_{11} cutoff V-values as a function of theaspect ratio ϵ calculated using MFDM and FEM. The agreement is better that 0.1% over the range covered

ϵ	$LP_{11}^{oe} : V_{co}$		$LP_{11}^{eo} : V_{co}$	
	MFDM	FEM	MFDM	FEM
0.5	1.81582	1.81593	2.66018	2.66024
0.6	1.88450	1.88460	2.49834	2.49824
0.7	1.95069	1.95076	2.37572	2.37558
0.8	2.01461	2.01464	2.27924	2.27910
0.9	2.07647	2.07646	2.20112	2.20102
1.0	2.13647	2.13641	2.13647	2.13641

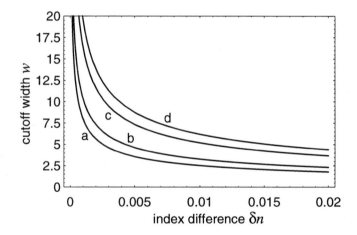

Figure 6.10 Cutoff width w_{co} for the square-core step-profile BCW as a function of the index difference $\delta n = n_{co} - n_{cl}$. These curves are calculated for $n_{co} = 1.45$ and $V_{co} = 2.136$. Labels a to d refer to wavelength values 0.63, 0.82, 1.3 and 1.55 μm, respectively.

A simple rational function approximation (Wolfram, 1988) can be derived using the data from Table 6.2. Furthermore, the cutoff V-value for the LP_{11}^{eo} mode with aspect ratio ϵ is identical to the LP_{11}^{oe} with aspect ratio $1/\epsilon$. Accordingly, the following rational function gives the cutoff V-value, with a maximum error of 0.1%, for the LP_{11}^{oe} over the range of aspect ratio valuesfrom 0.5 to 2:

$$V_{co} = \frac{1.4464 + 1.03\epsilon}{1 + 0.1594\epsilon} \tag{6.7}$$

There are few exact values of cutoff V-values available for rectangular- and square-core waveguides. For a step-profile rectangular core with $\epsilon = 1/2$, Eyges (Eyges *et al.*, 1979) used an expansion in cylindrical harmonics in the core and a corresponding expansion in harmonic functions of the two-dimensional Laplace equation in the cladding, together with point collocation along the core–cladding interface to give $V_{co} = 2.66$ for the LP_{11}^{eo} mode and $V_{co} = 1.87$ for the LP_{11}^{oe} mode. These points appear as the open dots on the left hand ordinate in Figure 6.9. For square cores, Goell (Goell, 1969) gives $V_{co} = 2.17$ compared with $V_{co} = 2.13$ obtained above.

6.3.3   Single-mode square-core BCWs

Using the V-value definition of (6.2) and the second-mode cutoff value $V_{co} = 2.136$ appearing in Table 6.2, we calculate a *cutoff width* w_{co}. This width is defined – in the case of the square-core BCW – as the maximum width (and height) of the core under which the waveguide operates in its monomode régime. Figure 6.10 plots the cutoff width w_{co} in terms of the index difference $\delta n = n_{co} - n_{cl}$ between the core and the cladding for four practical values of wavelengths. Details of the calculation are given in the caption.

REFERENCES

Ankiewicz, A. and Peng, G. D. (1992) Generalized Gaussian approximation for single-mode fibers. *IEEE Journal of Lightwave Technology*, **LT–10**, 22–27.

Eyges, L., Gianino, P. and Wintersteiner, P. (1979) Modes of dielectric waveguides of arbitrary cross sectional shape. *Journal of the Optical Society of America*, **69**, 1226–1235.

Goell, J. E. (1969) A circular-harmonic computer analysis of rectangular dielectric waveguides. *The Bell System Technical Journal*, **48**, 2133–2160.

Hewlett, S. J. and Ladouceur, F. (1995) Fourier Decomposition Method applied to mapped infinite domains: scalar analysis of dielectric waveguides down to modal cut-off. *IEEE Journal of Lightwave Technology*, **LT–13**, 375–383.

Press, W. H., Flannery, B. P., Teukolsky, S. A. and Vetterling, W. T. (1988) *Numerical recipes in C (The art of scientific computing)*. Cambridge: Cambridge University Press.

Sammut, R. A. (1982) A momentary look at noncircular monomodes fibres. *Electronics Letters*, **18**, 221–222.

Snyder, A. W. and Love, J. D. (1983) *Optical waveguide theory*. London: Chapman & Hall.

Wolfram, S. (1988) *Mathematica: a system for doing mathematics by computer*. Redwood city: Addison-Wesley.

7
Splicing loss

A major constraint on the choice of the cross-section geometry of BCWs is the requirement that the loss in splicing to standard single-mode fibres be minimal. The single-modedness of the fibre essentially requires the BCW to be single-moded as well. We show that minimal splice loss requires reasonable symmetry, and can be made virtually zero for square-core BCWs. Results are also presented for splice loss arising from geometrical mismatch, offset and tilt between the two cores, using the Fourier decomposition method (FDM). Approximate results based on the Gaussian approximation (GA) are also derived. We show that these analytical expressions are accurate approximations to the numerical results in the range of V-values of practical interest. The spectral variation of loss is also determined. For simplicity, we assume that throughout this chapter both the fibre and the BCW have the same step profile. We also address the effect of arbitrarily graded profiles on splice loss in Section 7.2.4.

7.1 MATHEMATICAL MODEL

Both the optical fibre and the BCW are assumed to be weakly guiding, so that it is sufficient to base the splice loss analysis on solutions of the scalar wave equation. Furthermore, the claddings of both waveguides are assumed to be sufficiently thick so that we can base the analysis on an unbounded cladding. This approximation introduces negligible error in the modal fields and propagation constants of practical fibres and BCWs with finite claddings. Accordingly, in an ideal splice, the fibre and waveguide abut one another with their axes parallel and coincident, and their faces are flush and normal to the axis, with no gap between them. The splice cross-section then has the geometry shown in Figure 7.1.

The claddings are assumed to have a common refractive index n_{cl}, and the cores have a common index n_{co}, i.e. the step profile. The effect of a variation in core profile between the fibre and the buried channel waveguide is discussed in Section 7.2.4 below.

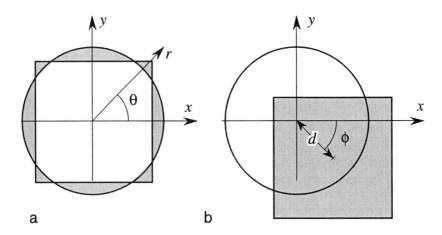

Figure 7.1 (a) Cross-section of the overlapped fibre and BCW cores, and (b) the common coordinate system for determining splice loss. Also shown is the lateral offset d of the BCW axis relative to the fibre axis and its inclination ϕ relative to the x-axis.

7.1.1   Excitation conditions

The fundamental mode of the fibre is incident on the BCW from $z < 0$, where the z-axis is coincident with the fibre and waveguide axes, and the splice coincides with the origin $z = 0$. Relative to polar coordinates (r, θ) in the cross-section, the incident electric field of the fibre is plane polarized with arbitrary polarization orientation and has the form

$$E_f = a_f e_f(r, \theta) e^{i\beta_f z} \tag{7.1}$$

where a_f is the modal amplitude, $e_f(r, \theta)$ the fundamental solution of the scalar wave equation for the transverse dependence and β_f the propagation constant. For the step profile, the fundamental solution of the scalar wave equation, normalized to unit amplitude on the core–cladding interface, is   (Snyder and Love, 1983, Table 14–3)

$$e_f = \frac{J_0(UR)}{J_0(U)}, \quad 0 \le R \le 1 \tag{7.2a}$$

$$e_f = \frac{K_0(WR)}{K_0(W)}, \quad 1 < R < \infty \tag{7.2b}$$

where $R = r/\rho_f$ is the normalized radial coordinate, ρ_f the fibre core radius, J_0 the Bessel function of the first kind, K_0 the modified Bessel function of the second kind, and the modal parameters U and W are defined by

$$U = \rho_f \left(k^2 n_{co}^2 - \beta^2\right)^{1/2}, \quad W = \rho_f \left(\beta^2 - k^2 n_{cl}^2\right)^{1/2} \tag{7.3a}$$

$$V_f = k\rho_f \left(n_{co}^2 - n_{cl}^2\right)^{1/2} = \left(U^2 + W^2\right)^{1/2} \tag{7.3b}$$

where V_f is the fibre parameter, and $k = 2\pi/\lambda$ the free-space wavenumber. The value of β_f depends on the smallest value of U satisfying the eigenvalue equation

$$U \frac{J_1(U)}{J_0(U)} = W \frac{K_1(W)}{K_0(W)} \tag{7.4}$$

and, because of the relations between V and the modal parameters in (7.3a), this value of U depends only on the value of the fibre V-parameter. However, β_f depends on V_f, ρ_f, k and n_{co}.

7.1.2 Reflected field

The mismatch in core geometries at the splice between the fibre and BCW partially reflects and partially transmits the incident fundamental-mode field of the fibre. The part of the incident modal field which is reflected at the splice excites the backward-propagating fundamental mode of the fibre and the continuum of backward-propagating radiation modes, which represents the portion of reflected power which is not guided (Snyder and Love, 1983, Chapter 25). In an exact analysis of splice loss, it would be necessary to quantify the amplitude of the reflected electric field at $z = 0$, because it is the total field at the interface, i.e. the sum of the incident and reflected electric fields, which determines the amplitude of the transmitted fundamental mode.

For present purposes, however, we can ignore the reflected field, because of the relatively small index difference betweeen n_{co} and n_{cl}, and assume that the total field at the interface is just the field of the incident fundamental mode of the fibre. We can justify this approximation by simple quantification. At each point on the interface, we assume that the incident electric field can be approximated by a local plane wave. The amplitude of the reflected electric field, E_r, is then given by the product of the incident field, E_f, and the Fresnel reflection coefficent (Snyder and Love, 1983, Section 20–2)

$$E_r = \frac{n_f - n_s}{n_s + n_f} E_f \tag{7.5}$$

where n_s and n_f are, respectively, the buried channel waveguide and fibre indices at that position on the interface. The reflected power is approximately proportional to the square of the integral of (7.5) over the infinite cross-section of the splice.

With reference to Figure 7.1a, it is clear that the only contributions to this integral come from those regions where n_s and n_f differ, i.e. from the eight regions between the intersection of the circular- and square-core cladding interfaces, in the case of identical step profiles. If we further assume that E_f is approximately constant over these eight regions because of the radial symmetry of the fibre, then the integral will be virtually zero if the areas of alternate shaded regions are equal, since the sign of $n_s - n_f$ alternates as θ increases. Cancellation of the positive and negative contributions to the integration from the eight regions in the intersection requires that the cross-sectional areas of the cores of the fibre and BCW be equal. This is the condition for minimal splice loss, as is confirmed numerically in Section 7.3.2.

7.1.3   Transmitted field

By analogy with the expression for the reflected field in (7.5), the transmitted field, E_t, which is the field exciting the buried channel waveguide has the form (Snyder and Love, 1983, Section 20–2)

$$E_t = \frac{2n_f}{n_s + n_f} E_f \tag{7.6}$$

The fibre and BCW have a common cladding index n_{cl}, and are both weakly guiding with only a slight variation between core and cladding indices, i.e. $n_f \approx n_s$ Hence the factor multiplying E_f differs little from unity over the whole cross-section, and the exciting field for the BCW can be taken to be the incident fundamental mode field of the fibre defined by (7.1), i.e. $E_t = E_f = a_f e_f$ at $z = 0$.

The transmitted field excites both the fundamental mode of the BCW and its radiation field. By analogy with (7.1), the fundamental BCW mode has the form

$$E_s = a_s e_s(x, y) e^{i\beta_s z} \tag{7.7}$$

where a_s is the modal amplitude, e_s is the fundamental-mode solution of the scalar wave equation for the transverse electric field dependence, relative to the cartesian coordinates in Figure 7.1b, and β_s is the propagation constant. The modal amplitude is proportional to the overlap integral between the fibre and BCW fundamental-mode fields on the interface (Snyder and Love, 1983, Section 20–5)

$$a_s = \frac{\displaystyle\int_{A_\infty} E_f e_s \mathrm{d}A}{\displaystyle\int_{A_\infty} e_s^2 \mathrm{d}A} = a_f \frac{\displaystyle\int_{A_\infty} e_f e_s \mathrm{d}A}{\displaystyle\int_{A_\infty} e_s^2 \mathrm{d}A} \tag{7.8}$$

where A_∞ is the infinite cross-section. The integral is evaluated at the splice $z = 0$, and e_f and e_s can be taken to be real, as both the fibre and BCW are assumed non-absorbing. Accordingly, the fraction of incident power transmitted to the fundamental mode of the BCW is given by

$$\frac{P_s}{P_f} = \left(\frac{a_s}{a_f}\right)^2 \frac{\int_{A_\infty} e_s^2 dA}{\int_{A_\infty} e_f^2 dA} = \frac{\left(\int_{A_\infty} e_f e_s dA\right)^2}{\int_{A_\infty} e_f^2 dA \int_{A_\infty} e_s^2 dA} \tag{7.9}$$

The fractional loss of power across the splice in dB is then calculated from

$$- 10 \log_{10}\left(\frac{P_s}{P_f}\right) \tag{7.10}$$

In the following section, we evaluate the integrals in (7.8) and (7.9) numerically using the FDM, while in Section 7.3 we introduce accurate analytical approximations based on the GA.

7.2   FDM ANALYSIS

The BCW fundamental mode amplitude given by (7.8) is proportional to the ratio of the overlap and normalization integrals, both of which involve the fundamental-mode field of the BCW. This ratio must be evaluated numerically, and the FDM offers the simplest numerical method. The BCW fundamental-mode field is expressed as a summation over the complete set of solutions of the scalar wave equation for the region enclosed by a perfectly-conducting square boundary with sides of length L, which is large compared with the core dimension ρ_s. Thus, by analogy with (5.3), we set

$$e_s(x,y) = \frac{2}{L} \sum_{m=1}^{N} \sum_{n=1}^{N} b_{mn} \sin\left(\frac{m\pi x}{L}\right) \sin\left(\frac{n\pi y}{L}\right) \tag{7.11}$$

relative to the coordinates in Figure 4.1, where the b_{mn} are constants to be determined, $m, n = 1, 2, \ldots, N$, and N is taken to be sufficiently large to ensure convergence.

Although the analytical form of the fibre fundamental-mode field is given explicitly by (7.2) in polar coordinates (r, θ), it is an inconvenient form for the purpose of evaluating the overlap integral in (7.8), because of the representation of the BCW fundamental-mode field in cartesian coordinates (x, y). It is simpler numerically to solve for the fibre modal field using the FDM technique by setting

$$e_f(x,y) = \frac{2}{L} \sum_{m=1}^{N} \sum_{n=1}^{N} c_{mn} \sin\left(\frac{m\pi x}{L}\right) \sin\left(\frac{n\pi y}{L}\right) \tag{7.12}$$

relative to the same boundary as for the BCW, but with coefficients c_{mn} to be determined.

The overlap integral is readily evaluated using (7.11),(7.12) and the orthog-

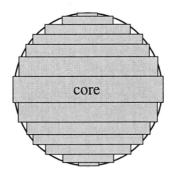

Figure 7.2 Set of rectangles approximating the circular fibre core for application of the FDM.

onality of the wave functions, leading to

$$\int_{A_\infty} e_f e_s dA \approx \int_0^L \int_0^L e_f e_s dx dy \tag{7.13a}$$

$$= \sum_{m=1}^{N} \sum_{n=1}^{N} b_{mn} c_{mn} \tag{7.13b}$$

Similarly, the normalization integrals for the BCW and fibre are, respectively

$$\int_{A_\infty} e_s^2 dA = \sum_{m=1}^{N} \sum_{n=1}^{N} b_{mn}^2, \quad \int_{A_\infty} e_f^2 dA = \sum_{m=1}^{N} \sum_{n=1}^{N} c_{mn}^2 \tag{7.14}$$

Hence, the fraction of fibre power entering the fundamental mode of the BCW follows from (7.9) as

$$\frac{P_s}{P_f} = \frac{\left(\sum_{m=1}^{N} \sum_{n=1}^{N} b_{mn} c_{mn}\right)^2}{\left(\sum_{m=1}^{N} \sum_{n=1}^{N} b_{mn}^2\right) \left(\sum_{m=1}^{N} \sum_{n=1}^{N} c_{mn}^2\right)} \tag{7.15}$$

and the corresponding splice loss is obtained by substituting into (7.10). The set of b_{mn} coefficients for the BCW is determined using the FDM, as described in Section 5.2, while the set of c_{mn} coefficients for the fibre is evaluated using exactly the same procedure but with the circular core of the fibre approximated by the set of rectangles illustrated in Figure 7.2.

Table 7.1 Numerical values of U, ρ_s/ρ_f, A_s/A_f, P_s/P_f and loss (dB) for the fibre–BCW splice as the ratio of core areas varies. The minimum loss case is highlighted

BCW V_s	BCW U_s	Ratio of radii ρ_s/ρ_f	Ratio of areas A_s/A_f	P_s/P_f	Loss [dB]
1.08675	1.02576	0.47250	0.28426	0.95198	0.21373
1.19543	1.09646	0.51975	0.34395	0.97241	0.12151
1.30410	1.15985	0.56700	0.40933	0.98323	0.07344
1.41278	1.21684	0.61425	0.48040	0.98911	0.04756
1.52145	1.26824	0.66150	0.55715	0.99268	0.03192
1.63013	1.31469	0.70875	0.63958	0.99518	0.02099
1.73880	1.35671	0.75600	0.72770	0.99708	0.01269
1.84748	1.39475	0.80325	0.82151	0.99848	0.00661
1.95615	1.42923	0.85050	0.92100	0.99929	0.00307
2.01049	1.44527	0.87413	0.97288	0.99945	0.00240
2.06483	1.46057	0.89775	1.02618	0.99941	0.00256
2.17350	1.48920	0.94500	1.13704	0.99872	0.00558
2.22784	1.50263	0.96863	1.19460	0.99804	0.00853
2.33651	1.52797	1.01588	1.31399	0.99599	0.01746
2.44519	1.55157	1.06313	1.43906	0.99299	0.03054
2.55387	1.57371	1.11038	1.56982	0.98906	0.04777
2.66254	1.59461	1.15763	1.70627	0.98421	0.06914
2.77122	1.61436	1.20488	1.84840	0.97846	0.09457
2.87989	1.63302	1.25213	1.99621	0.97185	0.12401
2.98857	1.65060	1.29938	2.14971	0.96441	0.15740
3.09724	1.66709	1.34663	2.30890	0.95616	0.19468
3.20592	1.68251	1.39388	2.47377	0.94716	0.23576

7.2.1   Variation with BCW core size

The first application of the FDM, described in the previous section, examines the variation of splice loss with the relative areas of the BCW and fibre cores, assuming identical step profiles and a common longitudinal axis, i.e no offset or tilt. The common core and cladding refractive indices are $n_{co} = 1.4514$ and $n_{cl} = 1.447$ (pure silica at a wavelength of 1.3μm), respectively, i.e. a relative index difference $\Delta = 0.003$. Assuming a source wavelength $\lambda = 1.3$ μm and fibre core radius $\rho_f = 4.233$ μm, the fibre V-value is $V_f = 2.3$, which ensures it is single-moded.

The results for the square-core BCW are presented in Table 7.1 showing the U-value, U_s, the ratio of the linear dimensions of the two cores ρ_s/ρ_f, the ratio of core areas $A_s/A_f = 4\rho_f^2/\pi\rho_s^2$, the fraction of power transmitted P_s/P_f, and

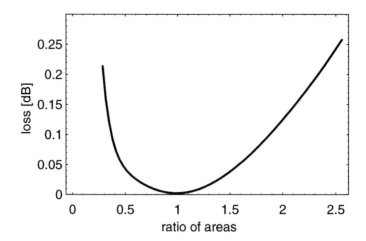

Figure 7.3 Fundamental-mode power loss in dB for the fibre/BCW splice, as a function of the ratio of BCW to fibre core areas, A_s/A_f.

the corresponding mismatch loss in dB as a function of the BCW V-value, V_s. The power loss is plotted against the ratio of fibre to BCW core areas in Figure 7.3.

The minimum loss occurs when the core areas of the two waveguides are virtually equal, as predicted in Section 7.1.2, i.e. when $\pi \rho_f^2 = 4\rho_s^2$ or $\rho_s/\rho_f = \sqrt{\pi}/2 \approx 0.886$. Clearly the actual loss is negligible over quite a wide range of BCW core size. If we impose a maximum acceptable loss of 0.1 dB on core mismatch, then the BCW core area can vary between 0.37 and 1.88 times the fibre core area. However, the BCW is not single-moded over all of this range. Table 6.2 of Section 6.3 shows that second-mode cutoff for the square-core step-profile BCW occurs when $V_s = 2.136$, where V_s is defined by (7.3b) with ρ_f replaced by ρ_s. Hence

$$V_s = \frac{\rho_s}{\rho_f} V_f = 2.3 \frac{\rho_s}{\rho_f} \tag{7.16}$$

Thus the BCW is single-moded only if $\rho_s/\rho_f < 0.926$; this includes the splice loss minimum at $\rho_s/\rho_f = 0.874$.

7.2.2   Variation with wavelength

For a practical splice, the loss minimum in Figure 7.3 is negligible; this suggests that the same conclusion should hold for a broad range of wavelengths, including both the 1.3 and 1.55 μm windows. To quantify this hypothesis, the half-length

Table 7.2 Variation of the loss minimum in Table 7.1 withwavelength

λ	V_f	U_f	V_s	U_s	P_s/P_f	Loss [dB]
1.20000	2.12308	1.57104	1.86462	1.39932	0.99943	0.00246
1.30000	2.30000	1.62284	2.02000	1.44783	0.99936	0.00277
1.40000	2.47692	1.66877	2.17538	1.49117	0.99928	0.00312
1.50000	2.65385	1.70982	2.33077	1.53014	0.99919	0.00351
1.60000	2.83077	1.74674	2.48615	1.56542	0.99910	0.00392

of the core boundary of the BCW was fixed at $\rho_s = 3.751$ μm, i.e. at the value of ρ_s/ρ_f corresponding to the loss minimum in Table 7.1, when $V_s = 2.02$. The wavelength of the source was then varied in the range 1.2 μm $< \lambda < 1.6$ μm, leading to the results given in Table 7.2 generated from the FDM through (7.15). Clearly, there is negligible spectral variation in splice loss over the whole range considered.

7.2.3   Effect of offset

Practical splicing of single-mode fibres and BCWs requires accurate alignment of the respective waveguide axes. This is complicated by the difference in the fibre and BCW geometries. Accordingly, it is useful to determine the sensitivity of splicing loss to offset, i.e. to the relative lateral displacement d of the two axes. To quantify this effect, we consider the two extreme orientations when the sides of the square core are either parallel to the x- and y-axes, as in Figure 7.1b, or are inclined at 45° to them.

The effect of the offset with parallel orientation is accounted for in the FDM by translating the axis of the BCW a distance d along the x-axis so that it is closer to one of the sides of the outer square boundary common to both the fibre and BCW. Similarly, the offset with 45° orientation is equivalent to translating the square core distance d along a diagonal of the bounding square. The losses for both orientations are listed in Table 7.3 as a function of the normalized offset d/ρ_f, using the same fibre and BCW parameters as used to generate Table 7.1 in the minimum mismatch loss configuration.

Firstly, we note that the actual loss is only marginally sensitive to orientation, the 45°-orientation splice loss being slightly greater than the parallel-orientation splice loss, up to $d/\rho_f = 0.66$, and then falling below it for larger values of d/ρ_f. The mean of the two is plotted against normalized offset in Figure 7.4. The offset loss in dB increases approximately exponentially with normalized offset. When the offset is equal to the fibre core radius, 4.233 μm, i.e. $d/\rho_f = 1$, loss is over

Table 7.3 Offset loss for the BCW with parallel or diagonal translation of the square core relative to the fibre core

d/ρ_f	U_s	Fractional loss	Loss [dB] $\phi = 0°$	Loss [dB] $\phi = 45°$
0.00000	1.45032	0.99968	0.00138	0.00138
0.20000	1.45027	0.96830	0.13988	0.14035
0.40000	1.45027	0.88152	0.54766	0.55120
0.60000	1.45043	0.75517	1.21953	1.22280
0.80000	1.45057	0.60824	2.15923	2.14934
0.84000	1.45059	0.57832	2.37830	2.36509
1.00000	1.45077	0.46258	3.34816	3.32793

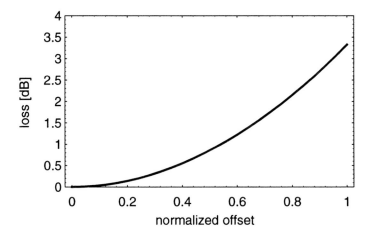

Figure 7.4 Splice loss plotted in dB as a function of normalized offset d/ρ_f.

50% (3.34 dB), while loss is only below 0.1 dB provided the offset is no more than 17% of the fibre core radius, i.e. 0.72 μm.

7.2.4   Effect of core index mismatch

In the analysis of splice loss undertaken so far, we have assumed that the fibre and BCW have identical step profiles. For practical telecommunications fibres and BCWs, this will not be the case in general, and, although the claddings may have a common refractive index (pure silica), the core profiles can be expected to differ quite markedly. There are two consequences of this difference. Firstly, this modifies the discussion of the reflected field in Section 7.1.2, and leads to a more complicated expression for the reflected power, which takes into account the detailed variation in the fibre core refractive-index profile. However, we can provide a simple upper bound on reflected power which requires knowledge only of the maximum difference between the fibre and BCW refractive-index profiles, n_f and n_s, respectively.

Suppose that the variation between n_f and n_s over the cross-section of the splice has a maximum value δ relative to the cladding index, i.e.

$$|n_f - n_s| < \delta n_{cl} \tag{7.17}$$

where $\delta \ll 1$. If we use this inequality in (7.5), then the bound on the reflected power P_r becomes

$$\frac{P_r}{P_f} < \left(\frac{n_f - n_s}{n_f + n_s}\right)^2 \approx \frac{\delta^2}{4} \tag{7.18}$$

since $n_f \approx n_s \approx n_{cl}$. For telecommunications fibres and BCWs, δ is at most 1%, so we deduce that reflected power will be at least 40 dB down. Accordingly, reflection from the interface is very small, and, consequently, the field exciting the BCW is well approximated by the incident fundamental-mode field of the fibre. The same conclusion holds for those splices which employ an index-matching liquid or epoxy to fill any possible voids between the fibre and the BCW.

The second consequence of the difference in profiles affects the excitation of the fundamental mode of the BCW, given by (7.8). The overlap integral depends on the product of the fibre and BCW fields, which must generally be calculated numerically, as in Section 7.2.1. However, as we show in the following section, the GA provides an accurate description of excitation, requiring knowledge only of the spot sizes of the fibre and BCW modes. This is discussed in further detail in Section 7.3.5.

7.3 GAUSSIAN APPROXIMATION ANALYSIS

Section 7.2 presented numerical results for splice loss. However, this approach requires extensive numerical analysis, and the dependence of loss on core mismatch or offset is not so readily apparent as in a more analytical approach. The latter can be provided by the GA, which was discussed in Section 4.4. As we show, this approximation leads to simple analytical expressions for loss, not just for mismatch and offset, but for tilt as well. The values of loss predicted by this approximation are close to the FDM values over the range of V-values of practical interest.

7.3.1   Fundamental mode

Within the Gaussian approximation, discussed in Section 4.4, we assume a Gaussian dependence for the transverse fields of the fundamental modes of both the fibre and BCW. If we define S_f and S_s to be the spot sizes for the fibre and square-core BCW normalized by the fibre core radius ρ_f, respectively, then relative to the fibre polar coordinates (r, θ), the corresponding fundamental-mode fields both have a purely radial Gaussian dependence

$$e_f = a_f \exp\left(-\frac{1}{2}\frac{R^2}{S_f^2}\right) \tag{7.19a}$$

$$e_s = a_s \exp\left(-\frac{1}{2}\frac{R^2}{S_s^2}\right) \tag{7.19b}$$

where $R = r/\rho_f$ is the radial coordinate normalized by the fibre core radius, and a_f and a_s are amplitude constants for the fibre and BCW fields, respectively.

The GA for the step-profile fibre leads to an analytical expression for the normalized spot size (Snyder and Love, 1983, Section 15–1)

$$S_f = \frac{1}{(2\log V_f)^{1/2}} \tag{7.20}$$

in terms of the fibre parameter (7.3a), and is valid for $V_f > 1$.

For the BCW, the GA for the fundamental mode of the square core leads to a transcendental equation for S_s given by (4.30), which is solved numerically. For reference, the U-values of the BCW in the range $V = 1.2$ to $V = 9.0$ are reproduced in the second column of Table 7.1. The value of U is in error with the FDM value for the BCW by less than 0.4% for V-values around 2.

The field on the interface between the fibre and the BCW was discussed in Sections 7.1 and 7.2.4, and is well approximated by the incident fundamental-mode field of the fibre. In other words, the exciting field for the BCW is given by (7.19a).

7.3.2   Variation with BCW core size

The integrand in the overlap integral in (7.8) is now the product of the fields in (7.19a) and (7.19b). Together with the normalization integral, this leads to the following closed-form expression for the fraction of transmitted power when there is a mismatch in core cross-sections (Snyder and Love, 1983, Table 20–1)

$$\frac{P_s}{P_f} = \frac{4(S_f S_s)^2}{(S_f^2 + S_s^2)^2} \tag{7.21}$$

Using the values of S_f and S_s from (7.20) and (4.30), respectively, leads, via (7.10), to the curve labelled 'GA' in Figure 7.5 for the loss due to the mismatch

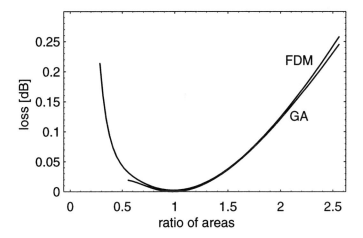

Figure 7.5 Comparison of the GA and FDM values of mismatch loss as a function of the ratio of BCW to fibre core areas A_s/A_f.

between the fibre and BCW cores, as a function of the ratio of BCW to fibre core areas. The FDM curve is reproduced from Figure 7.3. The two curves are virtually coincident over the range of validity of the GA. The loss minimum in the 'GA' curve is zero and occurs when the fibre and BCW spot sizes are identical, i.e. $S_f = S_s$ in (7.21). Clearly, the Gaussian approximation curve is a good approximation to mismatch loss over the range of normalized areas $0.7 < A_s/A_f < 2.5$.

7.3.3 Effect of offset

A simple analytical expression can also be derived for the fraction of the incident fundamental-mode power in the fibre which excites the fundamental mode of the BCW when the axes are offset distance d (Snyder and Love, 1983, Table 20–1)

$$\frac{P_s}{P_f} = \exp\left(\frac{-D^2}{S_f^2 + S_s^2}\right) \tag{7.22}$$

where $D = d/\rho_f$ is the offset normalized by the fibre core radius. The corresponding loss is calculated from (7.10), and is plotted in Figure 7.6 against the normalized offset, for parameter values corresponding to the loss minimum in Figure 7.5, i.e. $A_f = A_s$. The curve labelled 'FDM' is reproduced from Figure 7.4. For $d/\rho_f < 0.5$ the two loss curves are virtually coincident and at an offset of one core radius, i.e. $d/r_f = 1$, the difference is less than 15%.

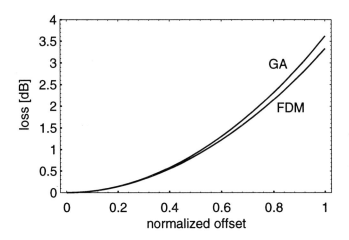

Figure 7.6 Comparison of the GA and FDM values of loss as a function of the normalized offset d/ρ_f.

If we take the product of (7.22) with (7.21), we obtain (Snyder and Love, 1983, Table 20–1)

$$\frac{P_s}{P_f} = \frac{4(S_f S_s)^2}{(S_f^2 + S_s^2)^2} \exp\left(\frac{-D^2}{S_f^2 + S_s^2}\right) \qquad (7.23)$$

which accounts for the effects of both core mismatch and offset.

7.3.4   Effect of tilt

If the axes of the fibre and the BCW are tilted relative to one another by angle θ, but are otherwise coincident where the two cores touch, the GA leads to the following expression for the fraction of fundamental-mode power transmitted (Snyder and Love, 1983, Table 20–1)

$$\frac{P_s}{P_f} = \exp\left[-\frac{(S_f S_s)^2}{(S_f^2 + S_s^2)^2}\,(k\rho_f n\theta)^2\right] \qquad (7.24)$$

where $k = 2\pi/\lambda$ is the free-space wavenumber, ρ_f is the fibre core radius, n is a representative value of the fibre index profile and the tilt angle θ is assumed small.

If we compare (7.22) and (7.24), we note that they are functionally identical if the normalized offset D in (7.22) is replaced by the normalized wavenumber

$k\rho_f n\theta S_f S_s$. In other words the offset loss in Figure 7.4 also provides the tilt loss under the transformation

$$D \rightarrow k\rho_f n\theta S_f S_s \tag{7.25}$$

Thus, for example, if the source wavelength is 1.3 μm, $n = 1.4514$, $\rho_f = 4.223$ μm and $V_f = 2.3$, we have $S_f = S_s = 0.876$ for zero loss due to core mismatch, then a tilt of 5°, or 0.087 radians leads to a loss of 3.54 dB, or over 50% power loss. Thus, only a tilt of 1° or 2° is tolerable in practical situations.

If we take the product of (7.21) with (7.24), then

$$\frac{P_s}{P_f} = \frac{4(S_f S_s)^2}{(S_f^2 + S_s^2)^2} \exp\left[-\frac{(S_f S_s)^2}{(S_f^2 + S_s^2)} (k\rho_f n\theta)^2\right] \tag{7.26}$$

accounts simultaneously for the effects of both core mismatch and tilt losses.

7.3.5   Effect of core mismatch

In Section 7.2.4, we discussed the effect of the difference between the core profiles of practical telecommunications fibres and BCWs, and deduced that the excitation efficiency of the BCW fundamental mode isproportional to the overlap integral between the fibre and BCW fundamental-mode fields. In general, this integral must be evaluated numerically, so it is not easy to anticipate its variation for a particular choice of fibre and BCW core profiles. Within the Gaussian approximation, however, the fraction of transmitted power, and hence the loss, is given by (7.21) in terms of the fibre and BCW normalized spot sizes. This result shows that loss will be minimal provided the spot sizes are sufficiently similar, even if the fibre and BCW refractive index profiles are distinctly different.

REFERENCES

Snyder, A. W. and Love, J. D. (1983) *Optical waveguide theory*. London: Chapman & Hall.

8
Surface roughness

While some modal attenuation is due to intrinsic material absorption, we find that material inhomogeneity and surface roughness also cause attenuation in fibres and planar waveguides through the scattering of guided optical power. However, the magnitude of this effect is very much larger for planar waveguides than for fibres. In the best single-mode BCWs currently reported, total scattering accounts for a loss of around 0.025 dB/cm at 1.3 μm (Adar *et al.*, 1991; Kawachi and Noda, 1990), whereas the Rayleigh scattering loss for the lowest-loss single-mode fibres operating in the 1.55 μm window is about 0.2 dB/km, i.e. approximately four orders of magnitude smaller. The reason for this large difference lies in the respective fabrication techniques.

8.1 MATERIAL INHOMOGENEITY

Fibres are normally drawn down from a heated, softened glass preform and, because of the tightly controlled drawing conditions, are very smooth and uniform along their length. The very low loss is mainly due to the fact that the scattering is small. It arises from imperfections in the drawn silica and its dopants, since absorption is negligible in the 1.3 and 1.55 μm windows. Similarly, the fabrication of BCWs by ion-exchange, ion-implantation and direct-write techniques also introduces nonuniformities into the host core materials, and hence scattering. However, as the detailed distribution of these nonuniformities is not readily available, it is difficult to predict the bulk scattering loss. Nevertheless, such loss can usually be reduced to acceptable levels through annealing, i.e. by heating the BCWs to sufficiently high temperature for a long enough period.

8.2 INTERFACE ROUGHNESS

In the PECVD fabrication of BCWs and devices, the buffer layer between the core and the silicon wafer, the core layer and the upper cladding layer are all produced by deposition. Accordingly, the two horizontal core–cladding interfaces might be expected to display some two-dimensional roughness, i.e. parallel to and transverse to the direction of propagation. Roughness can be

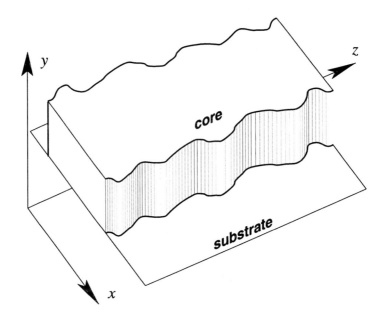

Figure 8.1 Schematic illustration of the substrate layer and the core layer after etching. Note the vertical striations representative of the masking and etching process. A cladding layer covers this structure to form a BCW.

measured accurately using the atomic force microscope (AFM), and is found to be so slight (of the order of a few nm), as to produce negligible scattering loss (Ladouceur *et al.*, 1992; Ladouceur *et al.*, 1994).

The masking and etching stages of BCWs, on the other hand, produce significant roughness, which appears predominantly as vertical striations, or corrugations on the two vertical side walls of the core. The two vertical core–cladding interfaces illustrated in Figure 8.1 are generated by etching away the core layer on either side of the mask width. These corrugations are due primarily to the inherent roughness along the edges of the mask defining the BCW, which tends to be reproduced by the etching process. As with the scattering due to material inhomogeneity, side-wall scattering can be significantly reduced by annealing.

Accordingly, in this chapter we develop the necessary analysis to predict the level of scattering loss from BCWs, as a function of roughness amplitude and distribution, as well as providing physical insight into the scattering process.

8.2.1   Radiation loss

In an optical waveguide, radiation occurs when there is any departure from the translational invariance, or cylindrical symmetry of the cross-section. To appreciate the physical origins of radiation caused by two-dimensional surface roughness, it is useful to consider two extreme situations.

Firstly, consider a waveguide where the roughness varies only in the azimuthal direction and not in the longitudinal direction. In the case of the BCW, the nominally square core–cladding interface would be deformed by slight irregularities running parallel to the waveguide axis. In other words, the BCW remains translationally invariant with a slight change in its cross-sectional geometry. Thus the fundamental mode propagates with a slight modification to its propagation constant and field, but there is no intrinsic radiation loss.

Secondly, consider a waveguide where roughness varies only in the longitudinal direction. This would consist of a set of irregularities distributed along its length. In this situation radiation loss depends critically on the correlation length of the roughness distribution, as we discuss in detail in Section 8.4. Thus it is clear that fundamental-mode power loss due to roughness depends predominantly on the longitudinal distribution of roughness, and the effect of azimuthal roughness is to introduce a slight modification to the modal propagation constant and fields. Accordingly, the following analysis is premised on a one-dimensional model of roughness.

8.2.2   Measurements of surface roughness

It is possible to measure the actual roughness of the BCW side walls very accurately (Ladouceur et al., 1992; Ladouceur et al., 1994). Figure 8.2 shows a typical roughness distribution, i.e. the deviation from perfect straightness, of the side wall of an etched BCW and clearly illustrates the randomness of the longitudinal roughness. These measurements were produced using an atomic force microscope (AFM) (Binnig et al., 1987), which consists of a cantilever-mounted tip that is scanned along the waveguide wall. The surface roughness deflects the tip, whose movement is monitored. The average displacement is the order of 0.05 μm, and when compared to the typical BCW width of 5 μm, corresponds approximately to a 1% variation.

8.3   MATHEMATICAL MODEL

Our model for calculating scattering from the fundamental mode, due to surface roughness of the BCW, is based on one-dimensional roughness in the direction of propagation. We also assume that the roughness on the two side walls of the

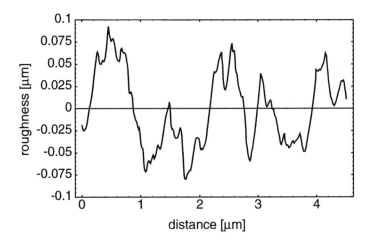

Figure 8.2 Typical experimental measurements of surface roughness along the vertical core–cladding interface of a BCW. Points denote the deviation from the mean core width of the BCW, denoted by the straight line.

interface is not correlated, so that the losses from each face are incoherent and can be summed independently to determine the total loss. Finally, in order to generate an analytical expression for the radiation loss, and delineate the role of the waveguide and roughness parameters, we assume a slab waveguide model for the pair of vertical interfaces. This approximation leads to a slight inaccuracy relative to using the exact BCW fields, but, like the Marcatili approximation of Section 4.2, gives a useful estimate of surface roughness radiation loss.

8.3.1   Characterization of surface roughness

The mathematical model for the calculation of roughness loss is illustrated in Figure 8.3, which shows the cross-section of a step-profile, nominally slab waveguide with slightly rough interfaces between the core and adjacent cladding layers. The unperturbed waveguide has core index n_{co}, cladding index n_{cl} and half-width $\rho(z)$. The mean value of the half-width along the waveguide is $\langle \rho(z) \rangle = \rho$, and, as discussed above, takes account of the roughness in the orthogonal, or y direction in the interfaces through averaging over y, i.e. $\rho(z) = \langle \rho(y, z) \rangle$.

We define a roughness function $f(z)$ as the local deviation of the perturbed surface from the perfect smooth interfaces. In terms of the index profile

$$n(x, z) = \begin{cases} n_{co} & \text{if } |x| < \rho + f(z) \\ n_{cl} & \text{if } |x| > \rho + f(z) \end{cases} \tag{8.1}$$

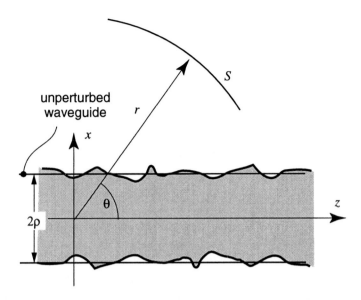

Figure 8.3 Roughness of the core–cladding interfaces defined in terms of displacement from perfectly smooth interfaces.

The power loss from the fundamental mode of the BCW due to the roughness is expected to be slight, so that perturbation methods can be used to quantify the loss (Marcuse, 1969). In the following analysis, the volume current method (VCM) is adopted (White and Snyder, 1977; Snyder and Love, 1983), which has been used to analyse radiation from corrugated fibres (Kusnetsov and Haus, 1983) and rough interfaces in slab waveguides (Lacey and Payne, 1990).

8.3.2   Spectral representation

The effect of surface roughness on propagation loss is more readily understood when the roughness function is described in terms of its spatial frequencies. Thus, we introduce the Fourier transform (FT) $\tilde{f}(\sigma)$ of $f(z)$ through the standard relations

$$\tilde{f}(\sigma) = \int_{-\infty}^{+\infty} f(z)\exp(-i\sigma z)\mathrm{d}z \tag{8.2a}$$

$$f(z) = \frac{1}{2\pi}\int_{-\infty}^{+\infty} \tilde{f}(\sigma)\exp(+i\sigma z)\mathrm{d}\sigma \tag{8.2b}$$

Given $\tilde{f}(\sigma)$, we define the spectral density of energyas $|\tilde{f}(\sigma)|^2$. This represents the 'intensity' of each component of spatial frequency σ of the roughness function.

8.3.3   Autocorrelation function

Another way to describe the roughness function and more specifically its random character is by introducing the autocorrelation function $C(u)$ defined by

$$C(u) = \lim_{L\to\infty} \frac{1}{2L} \int_{-L}^{+L} f(z)f(z+u)\mathrm{d}z \tag{8.3}$$

where $2L$ is the length of the waveguide. Thus $C(u)$ is a measure of the average correlation that exists between one point on the surface and another a distance u from it. It is defined in such a way that $C(0)$ gives the variance, δ_p^2, of the displacement from the smooth surface when the mean value of $f(z)$ is zero. The autocorrelation function is important as it is the physical quantity that can be measured in scattering experiments (Nélida et al., 1990; Fujii and Asakura, 1974), and conversely relating scattering loss to surface roughness requires the autocorrelation function $C(u)$, as shown below.

The crucial property is that the correlation $C(u)$ can be linked to the spatial frequencies present in the roughness function $f(z)$ through Wiener's hypothesis (Wiener, 1930; Bracewell, 1986), namely that for each roughness function $f(z)$ with correlation $C(u)$, there is a corresponding spectral density $S(\sigma)$ such that

$$C(u) = \int_{-\infty}^{+\infty} e^{i\sigma u} S(\sigma)\mathrm{d}\sigma \tag{8.4}$$

Under suitable conditions, this relation can be inverted through (8.2b) and leads to

$$S(\sigma) = \frac{1}{2\pi} \int_{-\infty}^{+\infty} e^{-i\sigma u} C(u)\mathrm{d}u \tag{8.5}$$

Equations 8.4 and 8.5 are known as Wiener–Khintchine's relations (Koopmans, 1974). Taking the FT of (8.3), as defined by (8.2b), shows that the FT of the correlation function $\tilde{C}(\sigma)$ is related to the FT of the roughness function $\tilde{f}(\sigma)$ by

$$\tilde{C}(\sigma) = \lim_{L\to\infty} \frac{1}{2L} \left|\tilde{f}(\sigma)\right|^2 \tag{8.6}$$

Furthermore, (8.5) can be rewritten as $S(\sigma) = \tilde{C}(\sigma)/2\pi$, which shows that

$$S(\sigma) = \lim_{L\to\infty} \frac{1}{2L} \frac{\left|\tilde{f}(\sigma)\right|^2}{2\pi} \tag{8.7}$$

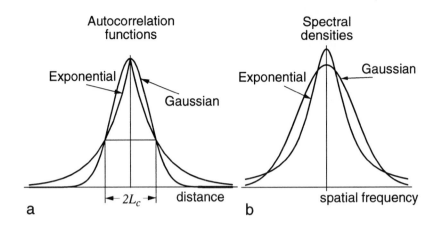

Figure 8.4 Schematic illustration of (a) the autocorrelation $C(u)$ and (b) the spectral density $S(\sigma)$ obtained via Wiener'stheorem.

Table 8.1 Exponential and Gaussian autocorrelation functions for surface roughness, together with their associated spectral densities, as deduced from Wiener's theorem. The third column gives the maximum of the spectral density when regarded as a function of the correlation length L_c.

Autocorrelation function $C(u)$	Spectral density $S(\sigma)$	Maximum $\max[S(\sigma)]$
$\delta_\rho^2 \exp\left(-\dfrac{\|u\|}{L_c}\right)$	$\dfrac{\delta_\rho^2 L_c}{\pi} \dfrac{1}{1 + L_c^2 \sigma^2}$	$L_c = 1/\sigma$
$\delta_\rho^2 \exp\left(-\dfrac{u^2}{L_c^2}\right)$	$\dfrac{\delta_\rho^2 L_c}{2\sqrt{\pi}} \exp\left(-\dfrac{\sigma^2 L_c^2}{4}\right)$	$L_c = \sqrt{2}/\sigma$

Typically the autocorrelation function $C(u)$ would be expected to exhibit a bell-shaped correlation between two points, and must decrease as their separation increases. This is illustrated in Figure 8.4.

There are two functions in particular, the exponential and the Gaussian, which have a bell-shaped variation and enable many analytical results to be derived (Ogilvy and Foster, 1989). The autocorrelation functions, correlation length L_c, corresponding spectral density functions and values of the correlation length at their maxima are presented in Table 8.1.

8.4   HEURISTIC APPROACH

The scattering of fundamental mode power due to surface roughness can be understood in terms of the loss induced by a regular sinusoidalcorrugation of period L, equivalent to a spatial frequency $\sigma = 2\pi/L$, i.e. by looking at the effect of one component frequency of the spectral distribution. This problem has been addressed by several authors using different methods (Snyder, 1970; Marcuse, 1975). There are three qualitative features common to the various results. If β denotes the propagation constant of the fundamental mode and k is the free-space wavenumber then:

- The only spatial frequencies of the corrugation leading to significant loss arebounded by

$$\beta - kn_{\mathrm{cl}} < \sigma < \beta + kn_{\mathrm{cl}} \tag{8.8}$$

- The radiation is scattered predominately at an angle θ relative to the waveguide axis, satisfying

$$\sigma = \beta - kn_{\mathrm{cl}} \cos\theta \tag{8.9}$$

- The radiation is proportional to the variance, δ_ρ^2, of the corrugation.

Physically, (8.8) and (8.9) show that significant radiation loss occurs when power is scattered into a direction which lies between forward-directed radiation, when $\sigma = \beta - kn_{\mathrm{cl}}$, $\theta = 0$, and backward-directed radiation, when $\sigma = \beta + kn_{\mathrm{cl}}$, $\theta = \pi$.

For surface roughness without any correlation, i.e. $L_c = 0$, all spatial frequencies of the roughness are uncorrelated and hence each frequency radiates independently of all the others. This leads to the standard result that, for uncorrelated signals, the total power radiated, P_{rad}, is just the sum of the powers radiated by each frequency. Hence

$$P_{\mathrm{rad}} \propto \int_{\beta - kn_{\mathrm{cl}}}^{\beta + kn_{\mathrm{cl}}} S(\sigma) \mathrm{d}\sigma \tag{8.10}$$

since $S(\sigma)$ is proportional to the spectral amplitude$|\tilde{f}(\sigma)|^2$. In Section 8.7, we show that, for partially correlated spatial frequencies, the superposition of powers still holds, but only for the main radiation lobe. Hence, recalling(8.9) and summing over all radiation angles we obtain

$$P_{\mathrm{rad}} \propto \int_0^\pi S(\beta - kn_{\mathrm{cl}} \cos\theta) \mathrm{d}\theta = \frac{1}{kn_{\mathrm{cl}}} \int_{\beta - kn_{\mathrm{cl}}}^{\beta + kn_{\mathrm{cl}}} S(\sigma) \left[1 - \left(\frac{\beta - \sigma}{kn_{\mathrm{cl}}} \right)^2 \right]^{-1/2} \mathrm{d}\sigma \tag{8.11}$$

Since the integrand is singular when $\sigma = \beta - kn_{\mathrm{cl}}$, and $S(\sigma)$ is a rapidly decreasing function, the integral can be well approximated by

$$\frac{S(\beta - kn_{\mathrm{cl}})}{kn_{\mathrm{cl}}} \int_{\beta - kn_{\mathrm{cl}}}^{\beta + kn_{\mathrm{cl}}} \left[1 - \left(\frac{\beta - \sigma}{kn_{\mathrm{cl}}} \right)^2 \right]^{-1/2} \mathrm{d}\sigma \simeq \pi S(\beta - kn_{\mathrm{cl}}) \tag{8.12}$$

Thus the scattering loss will be maximum when the correlation length satisfies

$$\frac{\mathrm{d}S(\beta - kn_{\mathrm{cl}})}{\mathrm{d}L_{\mathrm{c}}} = 0 \qquad (8.13)$$

This condition, together with the expressions given in Table 8.1, shows that for both the exponential and Gaussian autocorrelation functions, maximum scattering loss occurs when

$$L_{\mathrm{c}}^{\mathrm{max}} \simeq \frac{1}{\beta - kn_{\mathrm{cl}}} = \rho\sqrt{\frac{2}{\Delta}\frac{V}{W^2}} \qquad (8.14)$$

where V is the waveguide parameter, W the cladding modal parameter and Δ the relative index difference. For a step-index waveguide with $\Delta = 0.003$, $V = 2.0$ and $\rho = 5\ \mu$m, maximum scattering occurs when $L_{\mathrm{c}}^{\mathrm{max}} \simeq 100\ \mu$m.

The condition expressed by (8.14) has a simple physical interpretation. In a single-mode waveguide, the effect of the roughness can be described in terms of coupling from the fundamental mode to the radiation field. The length scale for this coupling is proportional to the beat length, $2\pi/(\beta - kn_{\mathrm{cl}})$, between the fundamental mode and the forward-directed radiation field. Hence we would expect roughness loss to be maximal when the correlation and coupling lengths are comparable, as expressed by (8.14).

8.5   VOLUME CURRENT METHOD

Here we formalize the discussion in the previous section and obtain quantitative results for roughness loss from the step-index slab waveguide.Firstly, we show how the VCM can be applied to any small perturbation of the waveguide, and then use the general loss expression for arbitrary perturbations to present an analytical expression for the attenuation of the fundamental mode. We emphasize that this analysis, although rigorous, requires experimentally measured values of the autocorrelation function in order for it to quantify the actual loss.

8.5.1   Equivalent current model

The VCM is based on the property that for a uniform dielectric waveguide, any perturbation is formally equivalent to a distribution of current dipoles within the unperturbed waveguide. Hence the perturbed waveguide can be treated as an antenna that radiates away part of the optical power transported by the modal field. If the perturbation is slight, the current density is given in terms of the field of the unperturbed waveguide, from which the far field and the loss are calculated.

8.5.2   Wave equation in the presence of currents

Consider the situation in Figure 8.3 for the symmetric step-profile slab wave-guide. As the current density is induced by the waveguide modal field, it has the same monochromatic time dependence and orientation as that of the TE-mode electric field component. Since the y-axis is perpendicular to the x–z-plane, we assume a current density distribution J_y orientated in the y–z-plane parallel to the y-axis. Accordingly we set

$$J_y = j_y e^{-i\omega t} \tag{8.15}$$

where j_y denotes the spatial distribution of J_y. It is then straightforward to show that the governing scalar equation containing the current source is (Snyder and Love, 1983)

$$\Delta e_y + k^2 n^2 e_y = -i\omega\mu_0 j_y \tag{8.16}$$

where Δ denotes the Laplace operator, μ_0 is the free-space permeability and ω the source frequency. The solution of this equation can be expressed in terms of the Green's function $G(\mathbf{r}, \mathbf{r}')$, which is the solution of the inhomogeneous equation

$$\Delta G(\mathbf{r}, \mathbf{r}') + k^2 n^2 G(\mathbf{r}, \mathbf{r}') = \delta(\mathbf{r}, \mathbf{r}') \tag{8.17}$$

where $\delta(\mathbf{r}, \mathbf{r}')$ is the Dirac delta function, $\mathbf{r} = (x, y, z)$ is the field position vector and $\mathbf{r}' = (x', y', z')$ is the position vector of the source. Using (8.17), the field e_y is then given by (Guenther and Lee, 1988)

$$e_y(\mathbf{r}) = -i\omega\mu_0 \int_{V'} j_y(\mathbf{r}) G(\mathbf{r}, \mathbf{r}') dV' \tag{8.18}$$

where V' is the volume occupied by the current densities. The exact determination of the Green's function from (8.17) is complicated because of the variation in the refractive index $n(\mathbf{r})$ across the waveguide. However, to lowest order within the weak guidance approximation the variation in index can be ignored, and the current density acts as if it was within a uniform medium with index equal to that of the cladding n_{cl}.

8.5.3   Free-space Green's function

Accordingly, we need the Green's function for a uniform medium with index n_{cl}. For the two-dimensional case it is given by (Bladel, 1964)

$$G(\mathbf{r}, \mathbf{r}') = \frac{i}{4} H_0^{(2)}(k n_{\mathrm{cl}} |\mathbf{r} - \mathbf{r}'|) \tag{8.19}$$

where $H_0^{(2)}$ is the Hankel function of the second kind of order zero, $\mathbf{r} = (y, z)$ and $\mathbf{r}' = (y', z')$. We require knowledge only of the far field in order to calculate the radiated power, for which the asymptotic form of (8.19) as $|\mathbf{r} - \mathbf{r}'| \to \infty$ is (Gradshteyn and Ryzhik, 1965)

$$G(\mathbf{r}, \mathbf{r}') \simeq \frac{i}{4} \left(\frac{2i}{\pi k n_{cl}} \right)^{1/2} \frac{\exp(ik n_{cl} |\mathbf{r} - \mathbf{r}'|)}{|\mathbf{r} - \mathbf{r}'|^{1/2}} \tag{8.20}$$

Furthermore, the position of the current source on the core–cladding interface is distant from the far field, i.e. $|\mathbf{r}'| \ll |\mathbf{r}|$, so we can expand $|\mathbf{r} - \mathbf{r}'|$ as

$$|\mathbf{r} - \mathbf{r}'| \simeq r - r' \cos \chi$$

where $r = |\mathbf{r}|$, $r' = |\mathbf{r}'|$ and χ is the angle between $\mathbf{r}$ and $\mathbf{r}'$ i.e $\cos \chi = \mathbf{r} \cdot \mathbf{r}' / r r'$. Thus (8.20) is well approximated by

$$G(\mathbf{r}, \mathbf{r}') \simeq \frac{i}{4} \left(\frac{2i}{\pi k n_{cl}} \right)^{1/2} \frac{\exp(ik n_{cl} r)}{r^{1/2}} \exp(ik n_{cl} r' \cos \chi) \tag{8.21}$$

Thus (8.18), together with the far-field Green's function, enables us to calculate the far field of the perturbed waveguide, provided the current density is prescribed.

8.6 RADIATED POWER

One can show, by rearranging the homogeneous scalar wave equation, that the effect of a slight perturbation on the uniform waveguide is formally equivalent to the incorporation of a 'fictitious' current density j_y within the unperturbed waveguide. It is defined by (Snyder and Love, 1983)

$$j_y = i \left(\frac{\epsilon_0}{\mu_0} \right)^{1/2} k(\bar{n}^2 - n^2) e_y \tag{8.22}$$

where k and ϵ_0 are, respectively, the free-space wavenumber and dielectric constant, $\bar{n}$ and n are, respectively, the refractive index profiles of the unperturbed and perturbed waveguides, and e_y is the scalar fundamental mode field of the unperturbed waveguides. By substituting (8.21) and (8.22) into (8.18), we obtain the far field as

$$e_y(\mathbf{r}) = \frac{\omega \mu_0}{4} \left(\frac{2i}{\pi k n_{cl}} \right)^{1/2} \frac{\exp(ik n_{cl} r)}{r^{1/2}} \int_{V'} j_y \exp(ik n_{cl} r' \cos \chi) dV' \tag{8.23}$$

The total power radiated away from the waveguide is then calculated using the integrated Poynting vector

$$P_{rad} = \frac{1}{2} \mathrm{Re} \left[\int_{S_\infty} (\mathbf{E} \times \mathbf{H}^*) \cdot \hat{\mathbf{r}} dS \right] \tag{8.24}$$

where S_∞ is the circle of infinite radius, $\mathbf{E}$ and $\mathbf{H}$ are the far-field electric and magnetic fields, respectively, $*$ denotes complex conjugate, and $\hat{\mathbf{r}}$ is the unit outward vector on the circle. At large distances from the waveguide, the radiation can be locally approximated by a plane wave, whence $\mathbf{H} = n_{\mathrm{cl}}(\epsilon_0/\mu_0)^{1/2}\hat{\mathbf{r}} \times \mathbf{E}$. With the help of (8.15), we obtain

$$P_{\mathrm{rad}} = \frac{n_{\mathrm{cl}}}{2}\left(\frac{\epsilon_0}{\mu_0}\right)^{1/2}\int_0^{2\pi}|e_y(\mathbf{r})|^2 r\,dr \qquad (8.25)$$

in terms of the field of (8.23).

8.6.1   Surface roughness as a current density

Equation (8.22) enables us to replace the surface roughness distribution, $f(z)$, by an effective current density. If we assume that the two interfaces of the slab waveguide in Figure 8.3 are uncorrelated, then the power scattered from both interfaces is simply twice that of the single interface, provided the roughness on each interface is stochastically identical. Conversely, if the two interfaces are perfectly correlated, then the far fields of each interface interfere possibly constructively, producing an upper limit of four times the loss from a simple interface. As discussed in 8.3.3, the roughness of the two interfaces is uncorrelated because of the nature of the lithographical and etching processes.

Accordingly, we need the equivalent current density for the upper interface only, and in (8.22) we set

$$\bar{n}^2 - n^2 = (n_{\mathrm{co}}^2 - n_{\mathrm{cl}}^2)f(z) \qquad (8.26)$$

with the fundamental-mode field expressed as

$$e_y(x, z) = a\phi(x)e^{i\beta z} \qquad (8.27)$$

where β is the propagation constant, $\phi(x)$ is the solution of the two-dimensional scalar wave equation, defined so that $\phi(\rho) = 1$, and a is the modal amplitude. Since the amplitude of the roughness distribution is small, we assume that the current density is located on the interface, and obtain

$$j_y = -ia(\epsilon_0/\mu_0)^{1/2}k(n_{\mathrm{co}}^2 - n_{\mathrm{cl}}^2)f(z)e^{i\beta z}\delta(x - \rho) \qquad (8.28)$$

where $\delta(x)$ is the Dirac delta function.

8.7   MODAL ATTENUATION

On substituting (8.28) into the far-field expression of (8.23) we find

$$e_y = \frac{iak^2(n_{\mathrm{co}}^2 - n_{\mathrm{cl}}^2)}{4}\left(\frac{2i}{\pi k n_{\mathrm{cl}}}\right)^{1/2}\exp(ikn_{\mathrm{cl}}z\sin\theta)\times$$

$$\int_{-L}^{+L}f(z)\exp\left(i(\beta - kn_{\mathrm{cl}}\cos\theta)z\right)dz \qquad (8.29)$$

where $2L$ is the length of the waveguide. By using the definition of theFourier transform (8.2), we deduce that

$$|e_y|^2 = \frac{a^2}{r} \frac{V^4}{\rho^4} \frac{1}{8\pi k n_{\mathrm{cl}}} \left| \tilde{f}(\beta - k n_{\mathrm{cl}} \cos \theta) \right|^2 \tag{8.30}$$

This contribution accounts for only one of the two interfaces and must be doubled for uncorrelated but equivalent roughness on both interfaces. Hence the total power radiated follows from (8.25) as

$$P_{\mathrm{rad}} = a^2 \frac{V^4}{\rho^4} \frac{1}{4k\pi} \left(\frac{\epsilon_0}{\mu_0} \right)^{1/2} \int_0^\pi \left| \tilde{f}(\beta - k n_{\mathrm{cl}} \cos \theta) \right|^2 d\theta \tag{8.31}$$

The attenuation of the fundamental mode power is expressed by (3.24)

$$P(z) = P(0) \exp(-\gamma z)$$

where $P(z)$ is the modal power at distance z along the waveguide and γ is the power attenuation coefficient, which is related to P_{rad} by (Snyder and Love, 1983, Section 22–5)

$$\gamma = \frac{P_{\mathrm{rad}}}{2LP(0)} = \frac{P_{\mathrm{rad}}}{2LNa^2} \tag{8.32}$$

where $N = \frac{1}{2} \int_{-\infty}^{+\infty} |(\mathbf{e} \times \mathbf{h}^*) \cdot \hat{\mathbf{z}}| \int dx$ is the normalization. Using Wiener's theorem we finally obtain

$$\gamma = \frac{1}{N} \frac{V^4}{\rho^4} \frac{1}{k} \left(\frac{\epsilon_0}{\mu_0} \right)^{1/2} \int_0^\pi S(\beta - k n_{\mathrm{cl}} \cos \theta) d\theta \tag{8.33}$$

in terms of the autocorrelation function.

8.7.1   Step-index slab waveguide

The fundamental-mode field of the slab waveguide has an exact analytical solution in the weak-guidance approximation, summarized in Snyder and Love, 1983, Table 12-2. Together with the condition that $\phi(x = \rho) = 1$, the normalization N is given by

$$N = \frac{\rho\beta}{2k} \left(\frac{\epsilon_0}{\mu_0} \right)^{1/2} \frac{V^2}{U^2} \frac{1+W}{W} \tag{8.34}$$

which on substituting into (8.33) gives

$$\gamma = \frac{1}{\rho^5 \beta} \frac{V^2 U^2 W}{1+W} \int_0^\pi S(\beta - k n_{\mathrm{cl}} \cos \theta) d\theta \tag{8.35}$$

for the power attenuation coefficient.

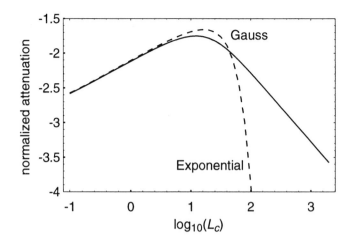

Figure 8.5 Normalized attenuation coefficient $\bar{\gamma}$ for the fundamental mode of the step index slab waveguide. Core and cladding indices are $n_{co} = 1.447$ and $n_{cl} = 1.444$, respectively, core half-width is $\rho = 4.443$ μm, and $V = 2.0$ at a wavelength of 1.3 μm. L_c is the normalized correlation length.

8.8   EVALUATION OF RESULTS

The normalized attenuation coefficient $\bar{\gamma} = \gamma \rho^3 / 2\delta_\rho^2$ for the step-index slab waveguide is plotted in Figure 8.5 as a function of the normalized correlation length, L_c/ρ, through numerical evaluation of (8.35) for both the exponential and Gaussian autocorrelation functions of Table 8.1. Both curves exhibit similar features, with a maximum value of attenuation occurring at a correlation-length value which is well approximated by (8.14), which predicts that $L_c^{max} \simeq 20\rho \simeq 100$ μm. For values of L_c around this maximum, the attenuation coefficient is given by

$$\gamma = \bar{\gamma} \frac{2}{\rho} \left(\frac{\delta_\rho}{\rho} \right)^2 \text{cm}^{-1} \tag{8.36}$$

where $\bar{\gamma}$ is the normalized attenuation from Figure 8.5. Hence the maximum loss possible due to surfaceroughness is strongly dependent on the ratio δ_ρ/ρ. If the loss is not to exceed 0.01 dB/cm, then the relative fluctuation of the core thickness δ_ρ/ρ should be kept below 1%.

REFERENCES

Adar, R., Shani, Y., Henry, C. H., Kistler, R. C., Blonder, G. E. and Olsson, N. A.

(1991) Measurement of very low-loss silica on silicon waveguides with a ring resonator. *Applied Physics Letters*, **58**, 444–445.

Binnig, G., Gerber, Ch., Stoll, E., Albrecht, T. R. and Quate, C. F. (1987) Atomic resolution with atomic force microscope. *Europhysics Letters*, **3**, 1281–1286.

Bladel, J. V. (1964) *Electromagnetic fields.* New York: McGraw-Hill.

Bracewell, R. N. (1986) *The Fourier transform and its applications, 2nd ed.* New York: McGraw-Hill.

Fujii, H. and Asakura, T. (1974) Effect of surface roughness on the statitical distribution of image speckle intensity. *Optics Communications*, **11**, 35–38.

Gradshteyn, I. S. and Ryzhik, I. M. (1965) *Table of integrals, series, and products.* London: Academic Press.

Guenther, R. B. and Lee, J. W. (1988) *Partial differential equations of mathematical physics and integral equations.* Englewood Cliffs, New Jersey: Prentice-Hall.

Kawachi, M. and Noda, J. (1990). Large scale integrated optic circuits using silica waveguides. Pages 101–104 of: *Proceedings OFS-7.*

Koopmans, L. H. (1974) *The spectral analysis of time series.* New York: Academic Press.

Kusnetsov, M. and Haus, H. A. (1983) Radiation loss in dielectric waveguides structures by the volume current method. *IEEE Journal of Quantum Electronics*, **QE–19**, 1505–1514.

Lacey, J. P. R. and Payne, F. P. (1990) Radiation loss from planar waveguides with random wall imperfections. *IEE Optoelectronics Part-J*, **137**, 282–288.

Ladouceur, F., Love, J. D. and Senden, T. J. (1992) Measurement of surface roughness in buried channel waveguides. *Electronics Letters*, **28**, 1321–1322.

Ladouceur, F., Love, J. D. and Senden, T. J. (1994) Effect of side wall roughness in buried channel waveguides. *IEE Optoelectronics Part-J*, **141**, 242–248 .

Marcuse, D. (1969) Radiation losses of dielectric waveguides in terms of the power spectrum of the wall distorsion function. *The Bell System Technical Journal*, **48**, 3233–3242.

Marcuse, D. (1975) Radiation losses of the HE_{11} mode of a fiber with sinusoidally perturbed core boundary. *Applied Optics*, **14**, 3021–3025.

Nélida, A., Russo, N. A., Bolognini, N. A., Sicre, E. E. and Garavaglia, M. (1990) Surface roughnesss measurement through a speckle method. *International Journal of Optoelectronics*, **5**, 389–395.

Ogilvy, J. A. and Foster, J. R. (1989) Rough surfaces: gaussian or exponential statistics. *Journal of Physics D: Applied Physics*, **22**, 1243–1251.

Snyder, A. W. (1970) Radiation losses due to variations of radius on dielectric or optical waveguide. *IEEE Transactions on Microwave Theory and Techniques*, **MTT–18**, 608–615.

Snyder, A. W. and Love, J. D. (1983) *Optical waveguide theory.* London: Chapman & Hall.

White, I. A. and Snyder, A. W. (1977) Radiation from dielectric optical waveguides. *Applied Optics*, **16**, 1470–1472.

Wiener, N. (1930) Generalised harmonic analysis. *Acta Mathematica*, **55**, 117–258.

9
Substrate leakage

The fabrication methods using the deposition and etching processes described in Chapter 2 are used to form planar waveguides and devices on a given substrate. As a commonly used substrate is the silicon wafer, it is important to isolate the guiding core region sufficiently far from the substrate, as silicon has a refractive index of 3.49 at a wavelength of 1.3 μm (Gray, 1972), which is significantly higher than that of the core and cladding silica-based materials, with refractive indices in the range 1.44–1.5. The higher index of the silicon gives rise to substrate leakage as the light propagating in the core will leak from the lower index buffer layer of silica into the higher-index silicon substrate through *optical tunnelling*, which is analogous to frustrated total internal reflection. In the silicon, the light is radiated away from the waveguide, as silicon is relatively non-absorbing in the near infrared.

In this chapter, we devise a simple model of substrate leakage and use it to quantify the leakage loss from the core through the lower cladding, or buffer layer to the substrate (Ladouceur, 1991). Intuitively, the leakage loss will decrease with increased buffer layer thickness; our model allows quantification of the minimum buffer layer thickness required to keep this leakage below a predetermined level. The model and results can also readily be adapted to quantify the minimal thickness of the upper cladding if it is coated with a higher-index protective layer, such as acrylate, which is absorbing in the near infrared. Although the calculation here is premised on the fundamental mode, it is readily generalized to higher-order modes.

9.1 MATHEMATICAL MODEL

The simplest mathematical model for quantifying tunnelling through the buffer layer to the substrate assumes that the square core cross-section in Figure 9.1 can be regarded as a section of a slab waveguide of infinite width in the y-direction and uniform cross-section 2ρ in the x-direction. It is sufficiently accurate for present calculations to assume that the cladding thickness above the core is infinitely thick, which help simplify the analysis. Accordingly, the cladding–core–cladding–substrate slab waveguide has the refractive-index profile shown in Figure 9.2.

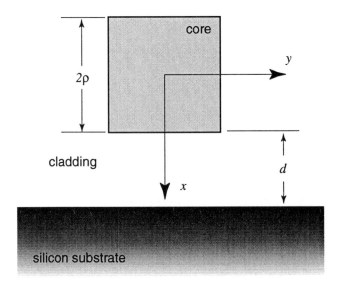

Figure 9.1 Cross-section of a BCW showing the substrate, orientation of the x- and y-axes used in the analysis and the substrate thickness d.

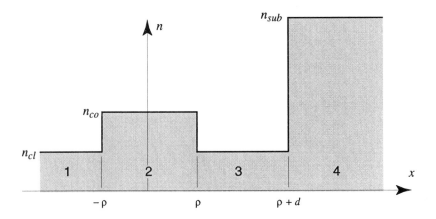

Figure 9.2 Refractive index profile for the slab waveguide model of the BCW and substrate.

While this model does not use the exact modal field distribution of the square-core waveguide, it has the advantage that it leads to a simple, explicit analytical solution, provided the leakage loss is not too large. As we show in Section 9.2, a simple modification can be made to the solution in order to take account of the difference between the slab and square-core field distributions.

9.1.1   Modal attenuation

Leakage of optical power from the guiding structure to the substrate results in attenuation of the fundamental mode of the BCW. This is accounted for analytically by allowing the propagation constant to become complex, i.e. $\beta = \beta^r + i\beta^i$, where superscripts r and i denote real and imaginary parts, respectively. The imaginary part β^i allows for modal attenuation along the BCW by introducing the factor $\exp(-\beta z)$ into the field expression. Thus, within weak-guidance, the scalar modal field $E(x, z)$ becomes

$$E(x, z) = a\phi(x)\exp(i\beta z) = a\phi(x)\exp(i\beta^r z - \beta^i z) \qquad (9.1)$$

where both the field $\phi(x)$ and the modal amplitude a are now complex, since they implicitly depend on β. The fundamental-mode power $P(z)$ at distance z along the guide is given by the integral of the z-directed component of the Poynting vector over the infinite waveguide cross-section. In the weak-guidance approximation, the modal power $P(z)$ at distance z along the waveguide is proportional to the integral of $|E(z)|^2$ over the infinite cross-section, and, therefore, proportional to $2\beta^i$. Hence $P(z)$ will decrease, as explained in Section 3.7, according to

$$P(z) = P(0)\exp(-2\beta^i z) = P(0)\exp(-\gamma z) \qquad (9.2)$$

where $\gamma = 2\beta^i$ is referred to as the power *attenuation coefficient*. The loss in dB is given by

$$-10\log\left[\frac{P(z)}{P(0)}\right] = 4.343\gamma \text{ dB/cm} \qquad (9.3)$$

when γ is expressed in cm^{-1}.

We determine the attenuation coefficient from the eigenvalue equation for the modes of the composite, leaky slab waveguide. As the propagation constant is complex, the eigenvalue equation will also be complex, and its fundamental root will determine both the real and imaginary parts of the propagation constant. However, as we shall show in Section 9.1.4, the imaginary part can be accurately determined analytically in terms of the real part, since $\beta^i \ll \beta^r$ for low-loss waveguides. The eigenvalue for the structure is derived by first determining the appropriate bounded fundamental-mode field solution of the scalar wave equation in each region of the structure, and then imposing the boundary conditions of continuity of these solutions and their first derivatives at each interface.

9.1.2 Modal fields

Within the weak-guidance approximation, the fundamental-mode field within each of the four regions of Figure 9.2 is a linear combination of the solutions of the one-dimensional scalar wave equation

$$\left[\frac{d^2}{dX^2} + \rho^2(k^2 n_i^2 - \beta^2)\right]\psi(X) = 0 \tag{9.4}$$

where $X = x/\rho$ is the normalized x-coordinate, k is the free-space wavenumber and n_i the refractive-index value in each region i. In terms of the effective index $n_{\text{eff}} = \beta/k$ for the fundamental mode, we have $n_{\text{eff}} > n_{\text{cl}}$ in the upper cladding region, $X < -1$, of Figure 9.2. The solution of (9.4) which is bounded as $X \to -\infty$ is

$$\psi(X) = Ae^{WX}, \quad -\infty < X < -1 \tag{9.5a}$$

where A is a constant and the complex modal parameters U and W have the same definitions as for non-absorbing waveguides, i.e.

$$W = \rho(\beta^2 - k^2 n_{\text{cl}}^2)^{1/2}, \quad U = \rho(k^2 n_{\text{co}}^2 - \beta^2)^{1/2} \tag{9.5b}$$

but are now in terms of the complex propagation constant β. In the core region 2, $n_{\text{eff}} < n_{\text{co}}$ so that the solution of (9.4) is the linear combination

$$\psi(X) = B\cos(UX) + C\sin(UX), \quad -1 < X < 1 \tag{9.5c}$$

where B and C are constants. In the lower cladding, or buffer region, $n_{\text{eff}} > n_{\text{cl}}$ and the solution of (9.4) is the linear combination

$$\psi(X) = Ee^{WX} + Fe^{-WX}, \quad 1 < X < 1 + D \tag{9.5d}$$

where E and F are constants and the normalized cladding thickness $D = d/\rho$. Finally, in the substrate region $n_{\text{eff}} < n_{\text{sub}}$, and, since the substrate is assumed infinitely thick and non-absorbing, we must satisfy the radiation condition, namely that the solution of (9.4) represent an outward-travelling wave for $X > D + 1$. Recalling the implicit time dependence $e^{-i\omega t}$, we set

$$\psi(X) = Ge^{iQX}, \quad 1 + D < X < \infty \tag{9.5e}$$

where G is a constant, and the complex modal parameter Q for the substrate region is defined as

$$Q = \rho(k^2 n_{\text{sub}}^2 - \beta^2)^{1/2} \tag{9.6a}$$

$$= \left(\frac{1 + 2\Delta}{2\Delta}\frac{n_{\text{sub}}^2 - n_{\text{cl}}^2}{n_{\text{cl}}^2}V^2 - W^2\right)^{1/2} \tag{9.6b}$$

in terms of the substrate index n_{sub}, the waveguide parameter V and the relative index difference Δ between the BCW core and cladding. The application of the weak-guidance approximation to the buffer–substrate interface with the relatively high difference in refractive index between silica and silicon might at first sight seem questionable. However, as we shall show: (i) the leakage loss is most sensitive to the buffer layer thickness, and (ii) the model is exact for the TE-polarized fundamental mode of the slab waveguide model, although slightly in error for the TM polarized mode because n_{sub} differs significantly from n_{cl}.

9.1.3   Eigenvalue equation

The continuity of $\phi(X)$ and $d\phi(X)/dX$ at the three interfaces $X = -1$, $X = 1$ and $X = D+1$ in Figure 9.2 leads to a set of six linear, homogeneous equations for the constants $A, \cdots, F$. These equations can be expressed as the product of a 6×6 matrix and a sixth-order column vector for $(A, \cdots, F)$. There is a non-trivial solution of these equations if and only if the determinant of the matrix vanishes, which determines the eigenvalue equation. Omitting straightforward algebraic details, we obtain

$$\frac{(W - U\tan U)(U + W\tan U)}{U + W\tan U + (W - U\tan U)\tan U} = \frac{W(W + iQ)e^{-2WD}}{(W + iQ)e^{-2WD} + W - iQ} \quad (9.7)$$

This equation is transcendental, so its exact solution would have to be obtained numerically. However, as we are only concerned with the solution corresponding to low leakage rates, i.e. $\beta^i \ll \beta^r$, we shall obtain an approximate, explicit solution in the following section.

9.1.4   Perturbation solution

If the buffer region 3 in Figure 9.2 were infinitelythick, (9.7) would reduce to the eigenvalue equation for the fundamental mode of the symmetric slab waveguide, i.e. (Snyder and Love, 1983, Chapter 12)

$$W = U\tan U \quad (9.8)$$

where W and U are now real. When the buffer thickness is finite but still large compared to the core width, we can seek a perturbation solution of (9.7) by setting

$$W = W^r + iW^i, \quad U = U^r + iU^i \quad (9.9)$$

where W^r and U^r denote the fundamental solution of (9.8), and $|W^i| \ll |W^r|$, $|U^i| \ll |U^r|$. Correct to first order in W^i and U^i, we then have

$$\tan U = \tan U^r + iU^i \sec^2 U^r \quad (9.10a)$$

$$W - U\tan U = -i\frac{U^i V^2}{U^r}\frac{W^r + 1}{W^r} \quad (9.10b)$$

Substituting into (9.7) and assuming $\exp(-2WD) \ll 1$, we deduce

$$U^i = -\frac{2(W^r)^3 U^r Q^r}{V^2[(W^r)^2 + (Q^r)^2]}\frac{e^{-2W^r D}}{W^r + 1} \quad (9.11)$$

By equating imaginary parts in the definition (9.5b) of U, the power attenuation coefficient is expressible as

$$\gamma = 2\beta^i = -\frac{2U^r U^i}{\rho^2 \beta^r} \tag{9.12}$$

On setting $\beta^r \approx k n_{co} \approx (V/\rho)(2\Delta)^{1/2}$, we finally obtain

$$\gamma = \frac{4}{\rho} \frac{(2\Delta)^{1/2}}{V^3} \frac{W^3 U^2 Q}{W^2 + Q^2} \frac{e^{-2WD}}{W + 1} \tag{9.13}$$

where we have dropped the superscript 'r', since all quantities on the right hand side are implicitly real. As discussed below, the values of U, V and W are taken to be those associated with the core–cladding BCW, and Q is then evaluated from (9.6).

9.2   LEAKAGE LOSS

The expression for the leakage loss, (9.13), shows that leakage to the substrate is most sensitive to the exponential term which, for given core–cladding and source parameters, depends on the normalized cladding thickness $D = d/\rho$. It is relatively insensitive to the variation in the modal and waveguide parameters with wavelength. The derivation of (9.13) is premised on the infinitely wide slab waveguide, and, therefore, would quantify leakage loss for that guide if the appropriate values U, V and W were used.

To account for the square core of the BCW, we retain the functional form of (9.13), and use the BCW values for U, W corresponding to the same value of V. Part of the justification for this choice of parameter values stems from an examination of the fraction of total fundamental-mode power that propagates within the core of the waveguide, denoted by η. For a slab waveguide, it has the analytical form (Snyder and Love, 1983, Table 12–2)

$$\eta = 1 - \left(\frac{U}{V}\right)^2 \frac{1}{W + 1} \tag{9.14}$$

while for the BCW it assumes the form

$$\eta = \text{erf}^2 \left(\frac{1}{S}\right) \tag{9.15}$$

in the GA, where S is the normalized spot size as defined in (4.21). Using these equations leads to the values listed in Table 9.1 and to the plots in Figure 9.3.

For a square-core waveguide with a $V = 2.0$, Table 9.1 gives $\eta \approx 0.79$. If we now calculate U and W from (9.8) for the slab waveguide using the same V-value, we find $\eta \approx 0.79$. In other words, the same fraction of modal power propagates in the claddings of both the BCW and slab models, provided we

Table 9.1 Fraction of fundamental-mode power η propagating in the core of the square-core BCW for the range of practical V-values

V	S	η
1.5	1.111080	0.635083
1.6	1.047020	0.677672
1.7	0.995682	0.713169
1.8	0.953397	0.743072
1.9	0.917824	0.768502
2.0	0.887380	0.790313
2.1	0.860959	0.809163
2.2	0.837758	0.825566
2.3	0.817181	0.839930
2.4	0.798774	0.852580
2.5	0.782186	0.863780

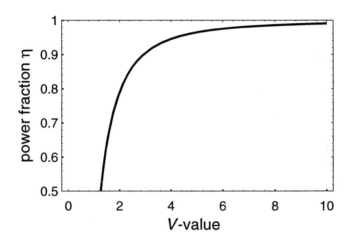

Figure 9.3 Plot of the fraction of BCW fundamental-mode power propagating in the core within the GA as a function of the waveguide parameter V.

use the same U-, V- and W-values. It should be noted that these values do not satisfy the eigenvalue equation (9.8) for the slab guide.

Hence, assuming a source wavelength of $\lambda = 1.3 \, \mu$m, core and cladding indices of $n_{\text{co}} = 1.45$ and $n_{\text{cl}} = 1.447$, respectively, a substrate index of $n_{\text{sub}} = 3.49$ and a core size of $\rho = 4.43873$ (corresponding to a V-value of $V = 2$), (9.6)

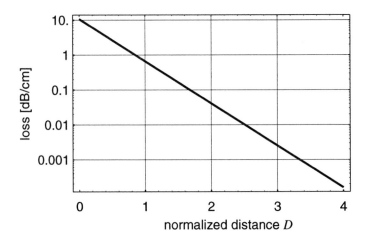

Figure 9.4 Substrate leakage for the fundamental mode of a square-core BCW on a silicon substrate as a function of the normalized cladding thickness d/ρ.

gives $Q = 60.4$. Substitution into (9.13) then leads to

$$\gamma = 10.68 \exp(-2.781D) \text{ dB/cm} \qquad (9.16)$$

which corresponds to the straight line in Figure 9.4 when plotted on a logarithmic scale. If we assume a maximum acceptable substrate loss of 0.01 dB/cm, then the minimum cladding thickness should not be less than 2.5ρ, i.e. at least comparable to the core thickness.

The above results actually provide a slight overestimate of the substrate leakage attenuation. While the BCW and slab waveguide model of leakage have the same fraction of modal power in their cladding, it is clear from the geometries of the respective core and substrate that the modal field of the BCW along the surface of the substrate parallel to the y-axis in Figure 9.1 will decrease on either side of $y = 0$, while it is constant in the slab model. Since leakage at each point on the surface of the substrate is proportional to the field intensity, the integrated loss for the BCW will be smaller than that for the slab model. Nevertheless, we emphasize that substrate loss is most sensitive to the thickness of the cladding between the core and substrate.

REFERENCES

Gray, D. E. (1972) *American Institute of Physics handbook*. New York: McGraw-Hill.

Ladouceur, F. (1991) *Buried channel waveguides and devices.* Ph.D. thesis, Australian National University.

Snyder, A. W. and Love, J. D. (1983) *Optical waveguide theory.* London: Chapman & Hall.

10

Bend loss

Bending of a single-mode waveguide leads to an attenuation of optical power in the fundamental mode as it propagates around the bend because of radiation loss from its modal field in the cladding. The magnitude of this loss has been established for both slab waveguides and fibres by a variety of methods, and an analytical formula for the attenuation has been derived, subject to plausible approximations consistent with practical situations(Marcatili, 1969; Snyder and Love, 1983).

In single-mode optical fibres, bending loss only becomes significant for bend radii less than a few centimetres, and is readily avoided in most practical situations where fibres are either embedded in relatively stiff cables or deliberately laid out in large-radius bends in fibre organizers. BCWs and devices, on the other hand, are frequently designed using very tight bends with radii of the order of millimetres, in order to make the most efficient use of the available substrate area, and therefore to maximize device yield and minimize production cost.

For example, in the BCW single-mode Y-junction discussed in Chapter 14, we show that the angle between the two output arms of the device must be sufficiently small, of the order of a degree or less, in order to minimize the excess loss of the device. However, if the two arms consist of straight, single-mode BCWs, then the Y-junction would have to be at least 1 cm long in order for the cores in the arms to be sufficiently well separated for splicing to two standard 125 μm diameter single-mode fibres with a core-to-core separation of 250 μm. Hence, in a concatenation of Y-junctions forming 1×4 and 1×8 splitters, each device would have to be several centimetres long. However, much more compact devices can be designed if bends are introduced into the arms. For example, as we will show in Chapter 14, the overall length of a single Y-junction satisfying the splicing separation requirement can be reduced to millimetres. Other devices, such as the couplers discussed in Chapter 12, also require bends in the BCW connections to minimize device length.

There are actually two contributions to bending loss which need to be taken into account. The first contribution is due to propagation along the bend of constant radius. This is the intrinsic pure bend loss and can be described by an attenuation coefficient which is independent of the length of the bend. Transition loss, on the other hand, arises from the relatively abrupt change from the infinite radius of curvature of the straight BCW to the finite radius or curvature of the bent waveguide. This change introduces a relative offset in the fundamental-mode fields of the straight and bent waveguides, leading to a misalignment of the modal fields and hence to transition loss.

In Chapter 4, we showed that the fundamental-mode field lines of a BCW, i.e. lines of constant electric field amplitude, are almost circular because of the four-fold symmetry of the square core cross-section. Furthermore, this field is well approximated by the GA of Section 4.4 within the range of V-values for practical single-mode operation. Accordingly, the development of the expressions for both the power attenuation coefficient for bend loss and the transition loss derived below, can be considerably simplified by using this approximation within and close to the core, with negligible loss of accuracy relative to using the exact bound-mode fields.

As we show, bend loss from a BCW is most sensitive to the bend radius and the relative index difference between the core and the cladding, and relatively insensitive to the core size and the source wavelength. We also compare pure bend loss with the transition loss and show how the latter may be readily avoided. The following Chapter 11 examines more sophisticated strategies for minimizing overall bend loss for a BCW linking two points in a planar circuit.

10.1   RADIATION CAUSTIC

The phase front of the fundamental mode of a straight BCW is a plane orthogonal to the direction of propagation, and the modal phase velocity parallel to the waveguide axis, ω/β, where ω is the souce frequency, has the same value everywhere on this plane. For the bent BCW of Figure 10.1 with constant radius R_c, it is intuitive that the phase plane rotates with constant angular velocity about the centre of curvature O in the plane of the bend. Hence the phase velocity in this plane must increase linearly with increasing distance from O in the plane of the bend, until, at a distance $r = r_{rad}$, it equals the speed of light in the cladding, c/n_{cl}, where c is the free-space speed of light. Beyond this position the field must necessarily radiate. Thus the position r_{rad} is sometimes known as the *radiation caustic*, since it represents the apparent origin of the radiation from the bend.

10.2   EQUIVALENT INDEX PROFILE

When a straight BCW is bent into an arc of constant radius, the effect of the bend on the modal field within and relatively close to the core can be modelled in terms of a straight BCW, but with a modified refractive index profile, referred to as the *equivalent index profile*. The equivalent profile can be deduced from

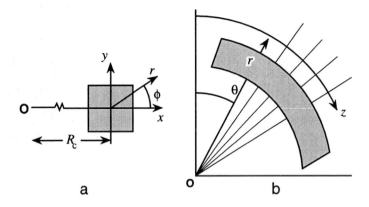

Figure 10.1 Two views of a uniformly-bent BCW: (a) cross-section of the bent BCW showing local cartesian and polar coordinates; (b) the plane of the bend showing the equal-phase plane rotating around the centre of curvature O.

simple physical considerations (Snyder and Love, 1983, Section 36–14). The planar phase front of the fundamental mode propagates around the bend in Figure 10.1 with a constant angular velocity Ω about O such that the phase $\Phi = \exp(i\kappa\theta)$ is constant on the planar front making an angle θ as shown with the beginning of the bend. We introduce a local propagation constant $\hat{\beta}$ relative to the axis of the bent BCW such that

$$\exp(i\kappa\theta) = \exp(i\hat{\beta}z) \tag{10.1}$$

where z is the distance measured along the axis of the waveguide from the beginning of the bend. Then, by geometry

$$z = [R_c + r\cos(\phi)]\theta \tag{10.2}$$

At the beginning of the bend the local propagation constant must equal that of the straight waveguide i.e. $\hat{\beta} = \beta$, and hence (10.1) and (10.2) give

$$\hat{\beta} = \frac{\beta R_c}{R_c + r\cos(\phi)} \approx \beta\left[1 - \frac{r}{R_c}\cos(\phi)\right] \tag{10.3}$$

since, by hypothesis, $R_c \gg r$. The propagation constant $\hat{\beta}$ can be considered as an effective propagation constant for a straight waveguide incorporating the effect of the bend. Hence, by replacing β by $\hat{\beta}$ in the scalar wave equation for the straight waveguide, and retaining terms correct to first order in r/R_c, we obtain

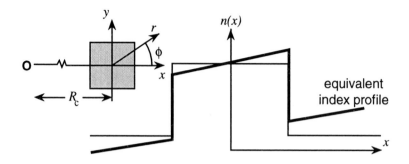

Figure 10.2 A bent waveguide can be regarded as a straight waveguide with an equivalent index profile given by (10.5), relative to the cross-section of a BCW, with the x-axis lying in the plane of the bend.

$$\Delta\psi(r,\theta) + (k^2 n_{\text{equ}}^2 - \beta^2)\psi(r,\theta) = 0 \qquad (10.4)$$

where n_{equ} is the equivalent index profile which, correct to first order, is given by the linear expression

$$n_{\text{equ}}^2 = n^2(\mathbf{r}) + 2n_{\text{co}}^2 \frac{r}{R_c}\cos(\phi) \qquad (10.5)$$

where the second term on the right hand side is small within and close to the core, since $r/R_c \ll 1$. This profile is illustrated in Figure 10.2.

An expression for the radius of the radiation caustic can be obtained from (10.4) and (10.5) by setting $\beta = kn_{\text{equ}}(r,\phi)$ and $r = r_{\text{rad}}$. Using the standard definitions, and $W = \rho(\beta^2 - k^2 n_{\text{cl}}^2)^{1/2}$, where ρ is the half-width of the square core, we find

$$\frac{r_{\text{rad}}}{\rho} = \frac{R_c \Delta}{\rho} \frac{W^2}{V^2 \cos(\phi)} \qquad (10.6)$$

Thus the position of the radiation caustic moves closer to the BCW as the bend radius decreases. This is indicative of increased radiation loss because of the larger field closer to the caustic. Thus we expect the bending loss to be small if the radiation caustic is far way from the waveguide axis i.e. if $r_{\text{rad}}/\rho \gg 1$, so that the scaled distance $R_c \Delta/\rho$ is the principle quantity involved. For a weakly guiding waveguide, with e.g. $\Delta = 0.0025$ and half-width $\rho = 5$ μm, we must have $R_c \gg 2$ mm in order to ensure low bending loss. A formal quantification of pure bend loss is derived in Section 10.5.

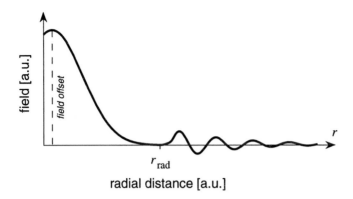

Figure 10.3 Schematic of the fundamental-mode field variation on the bent wave-guide, where the x-axis lies in the plane of the bend. The radiation caustic corresponds to the position where the effective index β/k and effective index profile $n_{\text{equ}}(x)$ are equal.

10.3   MODES OF BENT WAVEGUIDES

All modes of a bent BCW with an infinite cladding are necessarily leaky, including the fundamental mode, which has an electric field distribution in the cross-section of the bend shown schematically in Figure 10.3. Between the BCW axis and the radiation caustic, the field has approximately the same rapid monotonic decrease as the straight guide, but beyond the radiation caustic the field exhibits oscillations associated with the radiated power and a much slower decrease in amplitude.

We could quantify the bend loss by deriving the eigenvalue equation for the leaky modes of the bent BCW from the solutions of the scalar wave equation in the core and the cladding, using the boundary conditions on the electric field and its first derivatives at the core–cladding interface and imposing the radiation condition at infinity, i.e. the solution must represent an outward-travelling wave. Accordingly, the eigenvalue equation and its solutions for the propagation constant β become complex. By analogy with the discussion of the complex propagation constant for substrate leakage in Section 9.1.4, the real part of β determines the phase velocity of the mode along the bent waveguide axis, and (twice) the imaginary part determines the power attenuation coefficient along the bend.

However, there is no known coordinate system for bent BCWs which leads to analytical solutions of the eigenvalue equation. Even if we treat the equivalent index as a perturbation of the straight waveguide, the correction to the propagation constant of the fundamental mode is purely real (Garth, 1987), and no information about the loss loss can be obtained. However, by using the Volume Current Method (VCM) (Snyder and Love, 1983, Chapter 22), the radiation loss can be determined accurately, as we show in Section 10.5.

10.4   MODAL FIELD SHIFT

The perturbation analysis of bent waveguides is nevertheless important because it shows that an important effect of the curvature, as well as the modification to the real part of the propagation constant, is the offset of the modal field towards the outside of the bend (Marcatili, 1969; Arnaud, 1974; Heiblum and Harris, 1975). This offset accounts for the transient bend loss between a straight and curved section of the BCW. Intuitively, the transient loss increases with decreasing bend radius.

In Figure 10.1, the x-axis lies in the plane of the bend. The field shift is outward along this axis and can be determined using a variational procedure based on the trial function

$$\psi(x,y) = \left(1 + \epsilon\frac{x}{R_c}\right)\exp\left(-\frac{1}{2}\frac{x^2}{s_x^2}\right)\exp\left(-\frac{1}{2}\frac{y^2}{s_y^2}\right) = \phi(x)\exp\left(-\frac{1}{2}\frac{y^2}{s_y^2}\right)$$
(10.7)

where ϵ is a dimensionless parameter which quantifies the shift. When $\epsilon = 0$, the expression reduces to the Gaussian approximation of the fundamental mode field on the straight BCW, as discussed in Chapter 6. The s_x and s_y are known spot sizes, which are equal for the square-core cross-section. We use (10.7) as a trial function and substitute it into the following integral form of the scalar wave equation

$$\beta^2 = \frac{\int_{A_\infty} k^2 n_{equ}^2 \psi^2 - \left(\frac{\partial\psi}{\partial x}\right)^2 - \left(\frac{\partial\psi}{\partial y}\right)^2 dA}{\int_{A_\infty} \psi^2 dA}$$
(10.8)

where n_{equ} is the equivalent profile of (10.5). The value of ϵ is obtained from the extremum value of β. On performing the integration over y this reduces to

$$\beta^2 = \frac{\int_{-\infty}^{+\infty} k^2 n_{equ}^2 \phi^2 - \left(\frac{\partial\phi}{\partial x}\right)^2 dx}{\int_{-\infty}^{+\infty} \phi^2 dx} - \frac{1}{2a_y^2}$$
(10.9)

We drop the second term on the right hand side, since s_y is independent of ϵ. The field shift for the BCW is then precisely the same as that for the bent slab waveguide, because the right hand side of the equation depends only on x. The remaining integrals are readily evaluated, and on imposing the stationarity condition $\partial \beta^2 / \partial \epsilon = 0$ leads to

$$\frac{2s_x \rho k^2}{\sqrt{\pi}}(n_{\text{co}}^2 - n_{\text{cl}}^2)^2 e^{-\frac{\rho^2}{a_x^2}} + \frac{k^2 n_{\text{co}}^2 s_x^4}{R_c^2}\epsilon^2 + \epsilon - 2k^2 n_{\text{co}}^2 s_x^2 = 0 \qquad (10.10)$$

Since the ratios $s_x/R_c \ll 1$ and $\Delta \ll 1$, we neglect the first two terms, whence $\epsilon = 2(kn_{\text{co}}s_x)^2$. The shift in the maximum of the fundamental mode field, x_d, is determined from the trial expression of (10.7) by taking the derivative with respect to x. Hence, in terms of the waveguide parameter and relative index difference

$$x_d = \frac{V^2}{\Delta}\frac{s_x^4}{R_c \rho_x \rho_y} \qquad (10.11)$$

This expression is used in Section 10.7 to calculate the splicing loss between straight and bent sections of the BCW. Since the field distribution of the bent BCW is identical to that of the straight BCW, within the present approximation, the bent BCW field is identical to the field of the straight waveguide with the offset x_d. We can then use the analysis of splicing losses presented in Chapter 7.

10.5   PURE BEND LOSS

To determine pure bend loss for the BCW, we apply the VCM, which was used to quantify roughness loss in Chapter 8. It has an advantage over the leaky mode approach discussed above, in that it is readily applicable to the bent two-dimensional BCW. We will show that the exponential dependence in the attenuation coefficient, which dominates pure bend loss, is, to lowest order, independent of the shape of the core cross-section of the waveguide. Furthermore, the fundamental-mode field used in the calculation of the attenuation coefficient can be accurately represented by the GA, discussed in Chapter 4.

In the analysis of surface roughness in Chapter 8, the slight perturbation of the waveguide interface was modelled using an equivalent current distribution superposed on the unperturbed waveguide. A slightly different approach is adopted in the present application, whereby both the bent BCW and the fundamental-mode field are regarded as a perturbation of an infinite uniform medium with index equal to that of the cladding index n_{cl}. This perturbation is then expressed in terms of a current density distribution $\mathbf{J}$. Furthermore, the bend loss calculation is a three-dimensional problem, whereas the roughness loss calculation is only two-dimensional.

The detailed derivation of the corresponding attenuation coefficient due to bending of a fibre of circular cross-section is complicated and can be found elsewhere (Snyder and Love, 1983, Chapter 23). As the development given here is similar, it is sufficient to only give an outline for the bent BCW.

10.5.1   Volume current method

In the weak-guidance approximation, the polarization of the fundamental mode on a square-core BCW is arbitrary, as we showed in Section 3.5. Accordingly, we can ascribe a particular polarization, which, for simplicity in the analysis, is taken to be perpendicular to the plane of the bend, i.e. parallel to the y-axis in Figure 10.1. Then, by analogy with Section 8.6, the equivalent current density j_y is given by

$$j_y = i(\epsilon_0/\mu_0)^{1/2} k(n_{\rm cl}^2 - n^2) e_y \tag{10.12}$$

where e_y is the exact fundamental-mode field of the bent waveguide with profile n, polarized parallel to the y-axis. Thus j_y is non-zero only over the core of the bent waveguide and, as in Section 8.6, is well approximated by the field of the mode of the straight waveguide. The current density is then used asa source term in the vector wave equation of the vector potential $\mathbf{A}$

$$\left(\Delta + k^2 n^2\right) \mathbf{A} = -\mu_0 \mathbf{J} \tag{10.13}$$

The solution of this equation is given in terms of the free-space Green's function, and leads to the vector potential for the far field (Snyder and Love, 1983, Chapter 34)

$$\mathbf{A} = \frac{\mu_0}{4\pi s} \mathbf{M} \exp\left(iksn_{\rm cl}\right), \quad \mathbf{M} = \int_V \mathbf{J}(\mathbf{r}') \exp\left(-iks' n_{\rm cl} \cos\chi\right) dV' \tag{10.14}$$

where χ is the angle between the two position vectors, $\mathbf{r}$ and $\mathbf{r}'$, of norm s and s' respectively, and V' is the volume occupied by the current density. From a knowledge of the far-field vector potential $\mathbf{A}$, the power radiated is calculated according to

$$P_{\rm rad} = \frac{c^2 k^2 s^2 n_{\rm cl}}{2} \left(\frac{\epsilon_0}{\mu_0}\right)^{1/2} \int_{S_\infty} |\hat{\mathbf{r}} \times \mathbf{A}| \, ds \tag{10.15}$$

The calculation of $\mathbf{A}$ can then be carried out, assuming the bent guideforms a closed circular loop of radius R_c, and the attenuation coefficient γ is calculated from the equivalent expression to (8.32)

$$\gamma = \frac{P_{\rm rad}}{2\pi R_c P(0)}, \quad P(z) = P(0) \exp(-\gamma z) \tag{10.16}$$

Furthermore, we use the fact that the ratio of bend to core radii is large, so that the current density distribution can be approximated by a current $\mathbf{I}$ on the axis of the BCW, where

$$\mathbf{I} = \hat{\mathbf{n}} \int_{A_\infty} |\mathbf{J}| \, dA \tag{10.17}$$

Here $A_{\rm co}$ is the core cross-sectional area and $\hat{\mathbf{n}}$ a unit vector parallel to the direction of polarization of the field $\mathbf{E}$, i.e. perpendicular to the plane of the bend.

Accordingly, we model the bent waveguide as a thin-wire antenna radiating into homogeneous medium of index n_{cl}.

10.5.2   Equivalent line current

In order to obtain an explicit expression for the line current of (10.17), we define the waveguide parameter V and profile function f, respectively, relative to the cartesian coordinates of Figure 10.1a as

$$V = k(\rho_x \rho_y)^{1/2}(n_{\text{co}}^2 - n_{\text{cl}}^2)^{1/2} \tag{10.18}$$

$$n^2(X,Y) = n_{\text{co}}^2[1 - 2\Delta f(X,Y)], \quad X = \frac{x}{\rho_x}, \quad Y = \frac{y}{\rho_y} \tag{10.19}$$

where ρ_x and ρ_y are the half-widths of the core in the x- and y-directions, respectively. In order to simplify the notation, we also introduce the scale factor $\bar{\rho} = (\rho_x \rho_y)^{1/2}$. The magnitude of the scalar electric field $\mathbf{E}$ is given by

$$|\mathbf{E}| = a\phi(X,Y)e^{i\beta z} \tag{10.20}$$

so that the line current $\mathbf{I}$ can be expressed as

$$\mathbf{I} = I e^{i\beta z}\hat{\mathbf{n}} \tag{10.21a}$$

$$I = -ian_{\text{co}}\frac{V}{\bar{\rho}^2}\left(\frac{2\Delta\epsilon_0}{\mu_0}\right)^{1/2} \times$$
$$\int_{-\infty}^{+\infty}\int_{-\infty}^{+\infty}[1 - f(X,Y)]\,\phi(X,Y)\mathrm{d}X\mathrm{d}Y \tag{10.21b}$$

This expression gives the current density concentrated on the axis of the curved waveguide and when substituted into (10.14) and (10.15) leads to

$$\frac{P_{\text{rad}}}{P(0)} = \frac{I^2}{a^2 N}\frac{\sqrt{\pi R_c}}{16\bar{\rho}}\left(\frac{\epsilon_0}{\mu_0}\right)^{1/2}\frac{V^2}{W^{3/2}}\frac{1}{\Delta n_{\text{co}}}\exp\left(-\frac{4}{3}\frac{R_c}{\bar{\rho}}\frac{W^3}{V^2}\Delta\right) \tag{10.22}$$

Hence, using the definition of normalization

$$N = \frac{n_{\text{co}}\bar{\rho}^2}{2}\left(\frac{\epsilon_0}{\mu_0}\right)^{1/2}\int_{-\infty}^{+\infty}\int_{-\infty}^{+\infty}\phi^2(X,Y)\mathrm{d}X\mathrm{d}Y \tag{10.23}$$

we finally obtain

$$\gamma = \left(\frac{V^8}{64\pi R_c\bar{\rho}W^3}\right)^{1/2}\exp\left(-\frac{4}{3}\frac{R_c}{\bar{\rho}}\frac{W^3}{V^2}\Delta\right)$$
$$\times \frac{\left\{\int_{-\infty}^{+\infty}\int_{-\infty}^{+\infty}[1 - f(X,Y)]\phi(X,Y)\mathrm{d}X\mathrm{d}Y\right\}^2}{\int_{-\infty}^{+\infty}\int_{-\infty}^{+\infty}\phi^2(X,Y)\mathrm{d}X\mathrm{d}Y} \tag{10.24}$$

This expression holds for the fundamental mode of a bent waveguide of arbitrary core profile and cross-section. We note that the exponent, which dominates the variation of the attenuation coefficient, is independent of the detailed geometry of the core cross-section, but is sensitive to the normalized bend radius, R_c/ρ and the relative index difference Δ.

10.5.3   Square-core BCW

Application of (10.24) to the square-core, step-profile bent BCW requires numerical evaluation of the integrals, as there is no analytical solution of the scalar wave equation available. However, as we showed in Chapter 6, the GA is an accuraterepresentation of the fundamental-mode field of the straight BCW. Accordingly, if we substitute the approximation

$$\phi(X,Y) = \exp\left(-\frac{X^2 + Y^2}{2S^2}\right) \tag{10.25}$$

into (10.24), we obtain

$$\gamma R_c = \frac{\sqrt{\pi}}{2}\frac{V^4 S^2}{W^{3/2}}\mathrm{erf}^4\left(\frac{1}{\sqrt{2}S}\right)\left(\frac{R_c}{\rho}\right)^{1/2}\exp\left(-\frac{4}{3}\frac{R_c}{\rho}\frac{W^3}{V^2}\Delta\right) \tag{10.26}$$

where erf denotes the error function. The spot size S is determined from the eigenvalue equation (4.30)

$$\frac{2V^2}{\sqrt{\pi}}\exp\left(-\frac{1}{S^2}\right)\mathrm{erf}\left(\frac{1}{S}\right) = \frac{1}{S} \tag{10.27}$$

which is straightforward to solve numerically. Values of A are listed in Table 9.1.

10.6   EVALUATION OF RESULTS

10.6.1   Pure bend loss

To quantify pure bend loss, we use (10.26) and (10.27) to produce the curves in Figure 10.4 for the step-index, square-core BCW. Each curve relates the bend radius, normalized to the core half-width, to the profile height Δ, assuming a fixed loss of 0.01, 0.1 or 1 dB for right-angle (90°) bend. Parameter values used in the calculation are given in the caption.

As expected, the curves show that, for a given bend radius, loss decreases with increasing relative index difference, or, for a fixed relative index difference, loss decreases with increasing bend radius. For example, the 0.1 dB curve shows that if $\Delta = 0.003$, then $R_c \approx 4000\rho \approx 2$ cm for $\rho = 5\ \mu$m, whereas if $\Delta = 0.01$, then

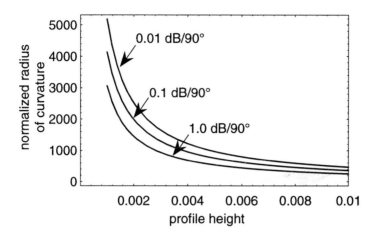

Figure 10.4 Curves of constant bending loss for a 90° bend as a function of the profile height Δ and the normalized radius of curvature R_c/ρ, for a square-core BCW with $V = 2.0$.

$R_c = 0.8$ mm for the same core size. In other words, increasing the relative index difference allows much tighter bends for a given loss, as guidance is stronger.

10.6.2 Effect of the finite cross-section

The attenuation coefficient of (10.26) is calculated from the linecurrent of (10.17). This approximation effectively superposes the whole core fundamental-mode field onto the axis of the BCW, i.e. it ignores the effect of the distribution of the field over the core. In the case of the step-profile fibre, it has been shown that taking into account the finite cross-section introduces a further factor multiplying the exponential termin (10.26). However, the correction due to core shape is found to be around 2% for the V-values in the practical monomode range 2.0 to 2.4.

Thus we can anticipate a similar correction for the BCW bending loss because ofthe similar symmetries of the core shapes. Numerical methods could be used instead of the GA to account for the finite core cross-section, but the increase in accuracy over using the GA would be small.

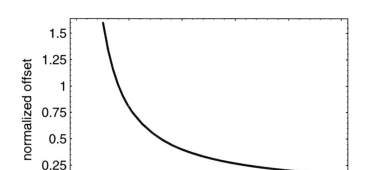

Figure 10.5 Normalized offset, r_d/ρ, calculated as a function of the normalized radius of curvature, R_c/ρ, for the step-index square-core BCW. This curve is obtained from (10.28) with $V = 2.0$, $\Delta = 0.003$ and $\rho = 4.0$ μm.

10.7 TRANSITION LOSS

The offset between the fundamental-mode fields of the straight and bent BCWs is given by (10.11). For the square-core BCW, this can be expressed in terms of the spot size, S, as

$$r_d = \frac{V^2}{\Delta}\frac{\rho^2 S^4}{R_c} \tag{10.28}$$

where S is the spot size determined from (10.27). The normalized offset r_d/ρ is plotted in Figure 10.5 as a function of the normalized bend radius R_c/ρ, showing the monotonic decrease with increasing bend radius. To quantify the transition loss L_T, we recall from Section 7.3 the expression for splice loss due to offset alone, when the two spot sizes are equal. In the present notation, this is equivalent to

$$L_T = -10\log\left[1 - \exp\left(-\frac{r_d^2}{2S^2}\right)\right] \tag{10.29}$$

The loss in dB is plotted as a function of the normalized bend radius in Figure 10.6. To compare the effects of pure bend loss and the transition loss, Figure 10.7 plots both curves for a BCW with the parameter values of Figure 10.6.

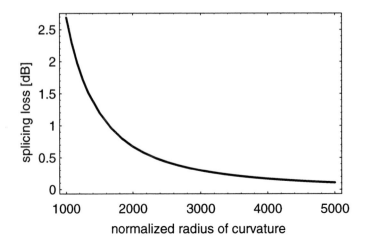

Figure 10.6 Transition loss, in dB, calculated as a function of the normalized radius of curvature for the step-index square-core BCW for an abrupt transition from a straight guide to a circular arc. This curve is obtained from (10.29) with $V = 2.0$, $\Delta = 0.003$ and $\rho = 4.0$ μm.

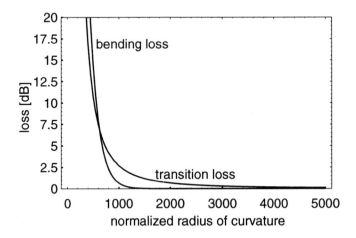

Figure 10.7 Transition loss in dB and bending loss in dB/90° as a function of the normalized radius of curvature for the step-index, square-core BCW calculated from (10.26) and (10.29) with $V = 2.0$, $\Delta = 0.003$ and $\rho = 4.0$ μm.

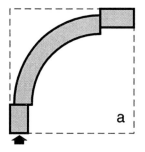 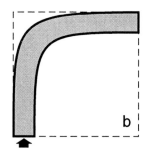

Figure 10.8 Two strategies to minimize transition loss: (a) offsetting the waveguides by the field shift of the bent waveguide; (b) continuously varying the waveguide curvature in an adiabatic fashion.

10.8   SUPPRESSION OF TRANSITION LOSS

Whilst it is not possible to avoid pure bending loss altogether, there are at least two strategies for minimizing the transition loss at the interface between the straight and bent BCWs. Transition loss occurs at the beginning of the bend because the fundamental-mode field is abruptly offset on the bend by r_d of (10.28). In one strategy, the bent waveguide is translated a distance r_d towards the centre of curvature, as illustrated in Figure 10.8a. In this case, the fundamental-mode fields on the straight and bent waveguides are aligned and transition loss is suppressed. The same strategy can also be applied at interfaces between two waveguides of opposite or different curvatures (Neumann, 1982).

The second strategy is based on the observation that the transient loss occurs because of the abrupt change in curvature between the straight and bent waveguides. If the curvature were to vary continuously and adiabatically, then the loss would be reduced. A transition bend with this property is illustrated in Figure 10.8b. This strategy is expanded upon in the following chapter.

REFERENCES

Arnaud, J. A. (1974) Transverse coupling in fiber optics Part III: Bending losses. *The Bell System Technical Journal*, **53**, 1379–1394.

Garth, S. J. (1987) Modes on a bent optical waveguide. *IEE Optoelectronics Part-J*, **134**, 221–229.

Heiblum, M. and Harris, J. H. (1975) Analysis of curved optical waveguides by conformal transformation. *IEEE Journal of Quantum Electronics*, **QE–11**, 75–83.

Marcatili, E. A. J. (1969) Bends in optical dielectric guides. *The Bell System Technical Journal*, **48**, 2103–2132.

Neumann, E. G. (1982) Curved dielectric optical waveguides with reduced transition losses. *IEEE Proceedings Part-H*, **129**, 278.

Snyder, A. W. and Love, J. D. (1983) *Optical waveguide theory*. London: Chapman & Hall.

11
Optimal path design

When designing a complex optical circuit incorporating a variety of devices suchas Y-junctions, X-junctions, couplers, wavelength multiplexers etc., one is confronted with the problem of linking each pair of input and output ports between successive devices by a waveguide. Since the position, orientation and curvature of each port within the plane of the circuit is in general arbitrary, the path followed by each waveguide is traditionally constructed using a concatenation of short segments of straight lines and arcs of circles, along which the cross-section and index profile of the waveguide remains unchanged. One reason for using these particular shapes is that pure bend loss has been calculated in detail for those. A second is that most mask-generating software, originating from micro-electronics, already caters for these particular shapes.

Nevertheless, it is intuitive, as far as bend loss is concerned, that the optimal path for linking two ports on an optical circuit is not necessarily one of these or a concatenation of them. Moreover, in many circumstances, this limited set forces the circuit designer to use it in an arbitrary and cumbersome manner. For example, the design of a low-loss S-bend which satisfies the geometrical constraints imposed by the ports at either end, and simultaneously minimizes bend loss, is a time-consuming process using that set.

Accordingly, there are two aims in this chapter. Firstly, we introduce a general family of curves that are not geometrically constraining, so that they can be optimized for both bending and transition loss (Ladouceur and Labeye, 1995). This family will link two arbitrary ports of an optical circuit, satisfying orientation and curvature conditions at both ports, and still leaving one free parameter which can be chosen to minimize bending loss. Secondly, we introduce a loss-minimizing procedure, based on a continuous variation of the core cross-section as the curvature varies, which, when used in conjunction with this general family of curves, leads to a substantial reduction in bend loss when compared to a concatenation of line segments and arcs of circles.

11.1 PLANAR MODEL

In Chapter 10, we showed that there are two types of bend loss which occur when connecting two ports in an optical circuit, namely pure bend loss and transition loss. In the following discussion, we use a single-mode, one-dimensional, slab model with a step profile to illustrate the basic concepts. This simplification is justifiable because both bend loss and transition loss depend predominantly on the BCW geometry in the plane of the bend, and are only slightly modified by the BCW geometry orthogonal to this plane.

The latter modification can be incorporated into the one-dimensional slab model, a priori, by using the effective index method (EIM) of Section 4.3, to accommodate the effect of the second cross-sectional dimension. This strategy can be applied to cross-sections other than that of the square or rectangular-core BCW. For example, consider the partially-etched, step-profile planar waveguide with the cross-section illustrated in Figure 11.1. Using the EIM, the fundamental mode of this asymmetric two-dimensional waveguide at a wavelength of 1.55 μm is equivalent to the fundamental mode of a symmetric slab waveguide with core index 1.471, cladding index 1.468 and core half-width 3.25 μm.

In the following, the slab waveguide half-width is designated by ρ, the core index by n_{co} and the cladding index by n_{cl}. The profile height is defined as

$$\Delta = \frac{n_{co}^2 - n_{cl}^2}{2n_{co}^2} \approx \frac{n_{co} - n_{cl}}{n_{co}} \tag{11.1}$$

for the weak-guidance approximation. The degree of guidance, or V-value, of the slab waveguide is given by

$$V = k\rho n_{co}\sqrt{2\Delta} \tag{11.2}$$

where k is the free-space wavenumber. The fundamental mode's effective index, n_{eff}, for the TE polarization can be calculated from the eigenvalue equation

$$W = U \tan U \tag{11.3}$$

where U and W are the core and cladding normalized eigenvalues defined by

$$U = k\rho\sqrt{n_{co}^2 - n_{eff}^2}, \quad W = k\rho\sqrt{n_{eff}^2 - n_{cl}^2} \tag{11.4}$$

Using these definitions, the slab waveguide is single-moded if $V < \pi/2$(Snyder and Love, 1983). The theory presented below is limited to the weak-guidance approximation but, since the physical concepts involved are independent of this approximation, our approach can be readily generalized.

11.2   PURE BEND AND TRANSITION LOSSES

11.2.1   Pure bend loss

We derived an approximate quantitative expression for pure bend loss for the square-core BCW in Section 10.5. A similar derivation can be used for the bent

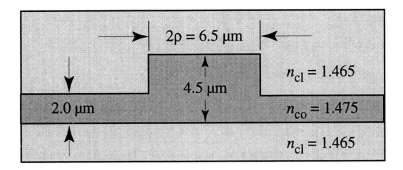

Figure 11.1 Cross-section, dimensions and indices of the partially etched rib waveguide used for simulation and case study.

slab waveguide, which leads to the following expression for the attenuation coefficient (Marcuse, 1973)

$$\gamma = \frac{1}{k n_{\text{eff}} \rho^2} \frac{U^2 W^2}{1 + W} e^{2W} \exp\left(-\frac{4}{3} \frac{W^3 \Delta}{V^2} \frac{R_c}{\rho}\right) \tag{11.5}$$

which exhibits the exponential decrease in bend loss with decreasing curvature $\kappa = 1/R_c$. It is more important for the present analysis that the loss decreases as the waveguide half-width ρ increases. For example, if $V = 1$ and ρ increases by 25%, then W increases from 0.67 to 0.93, and the expression $W^3/V^2\rho$ in the exponent increases by about 33%, with a commensurate decrease in bend loss. Thus, for example, if the waveguide becomes wider at an appropriate rate as the bend radius decreases, the net bend loss will be approximately unchanged and the resulting bend will be tighter. However, there are limits to this trade-off, which are imposed by the requirements that the waveguide remain single-moded, and the curvature changes slowly enough so that the local fundamental mode model remains valid throughout the bend.

11.2.2   Transition loss

A second consequence of bending is that it shifts the modal field slightly outwards from the centre of curvature in the plane of the bend, as was discussed for the bent BCW in Section 10.2. This effect can be understood in terms of the equivalent refractive index profile for the bend, where a linear correction is superposed on the step profile. This correctioncauses the modal field to be shifted outward towards the region of higher refractive index. This shift was calculated for the BCW using a variational approach in Section 10.2. A similar derivation for the slab waveguide leads to an offset given by

$$x_d = \frac{V^2}{\Delta} \frac{s_x^4}{R_c \rho^2} \tag{11.6}$$

where s_x is the spot size of the fundamental mode field. The latter is expressed as

$$\psi(x) = \psi_0 \exp\left(-\frac{1}{2}\frac{x^2}{s_x^2}\right) \tag{11.7}$$

within the Gaussian approximation, and the spot size s_x is calculated numerically from the one-dimensional analogue of (4.30)

$$\exp\left(\frac{1}{S_x^2}\right) = \frac{2S_x V^2}{\sqrt{\pi}}, \quad S_x = \frac{s_x}{\rho} \tag{11.8}$$

Like the corresponding expression for the BCW, this representation for the fundamental mode of the slab waveguide is accurate over the practical range of V-values for single-mode operation.

The transition loss is the loss in power of the fundamental mode due to the abrupt change in curvature at the interface between the straight and bent slab waveguides. It can be quantified using the overlap integral in a similar manner to that in Section 10.2.

11.3   TRANSITION LOSS MINIMIZATION

11.3.1   Core offset

An established strategy for essentially eliminating transition loss between a straight waveguide and a bent waveguide of constant curvature was discussed in Section 10.8. It requires the curved waveguide core axis to be offset relative to the straight waveguide axis at the beginning of the bend by a distance equal to the mode offset due to the curvature (Neumann, 1982), as illustrated in Figure 11.2a.

11.3.2   Core widening

Another strategy that can be used to reduce pure bend loss is based on the decrease of pure bend loss by widening of the waveguide core over the length of the constant radius bend, as indicated in Figure 11.2b. Here, the square core of the curved BCW would be widened abruptly at thejunctions with the straight waveguide. This strategy must be used inconjunction with an offset core to be of practical value in reducing overall loss (Pennings, 1990).

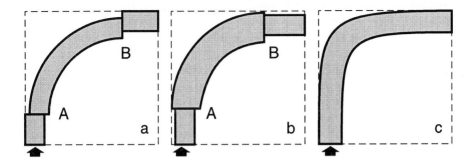

Figure 11.2 Three strategies to reduce total bend loss: (a) offset between the straight and bent waveguide axes; (b) offset of the straight and bent waveguide axes, combined with widening; (c) continuously varying the waveguide curvature from straight to bent in an adiabatic fashion.

11.3.3   Adiabatic curvature change

An alternative approach to the abrupt core offset of Figure 11.2a is to employ a continuous variation of curvature from zero, for the straight waveguide, to the desired value for the bent waveguide. To be effective, this approach requires that the fundamental mode propagates approximately adiabatically along the transition region. Adiabatic, in this situation, can be interpreted as meaning that the power attenuation for a particular value of curvature is just the pure bend loss value for that curvature. Such a strategy is illustrated schematically in Figure 11.2c.

11.4   LOW BEND-LOSS PATHS

11.4.1   Concatenated paths

Most current commercial mask-production software limits the design of waveguide paths to a concatenation of straight lines and arcs of a circle. Thus, within this limitation, the discussion of Section 11.2 would require (i) the introduction of offsets at a curvature discontinuity, i.e. straight to curved waveguides or from one arc to another of different radii, as in the case of the S-bend, and (ii) the use of a wider core for the curved sections in order to minimize total bend loss.

These two effects, offset and core widening, can be analysed using the formulae presented in Section 11.2, but optimizing them simultaneously can be tedious. Consider, for example, the 90° arc of the circle shown in Figure 11.2a.

The total loss is composed of transition losses at cross-sections A and B, together with pure bend loss along the arc. Widening the bent waveguide reduces pure bend loss but introduces mode mismatch at A and B, notwithstanding the offsets. This mismatch in turn limits the scope for core widening. In addition, the offset needed to minimize the mode mismatch depends on the widening, resulting in a two-parameter minimization procedure.

This calculation can be simplified using the Gaussian approximation with results given by (11.5) and (11.6), but, in practice, it turns out that the modes of both structures need to be calculated numerically to obtain acceptably accurate results. Furthermore, a real path is likely to be even more complex, involving several curved and straight sections. The total transition and bend loss for the optimal waveguide path can become very complicated, because the optimization at one discontinuity may depend on the precise location of the preceeding and following ones.

11.4.2   Continuous paths

The concatenation of discrete lengths of straight and curved waveguides discussed in the previous section will always have a finite total bend loss, even in the optimal design. Intuitively, a continuously-varying curved path based on the same underlying concepts should lead to a lower-loss optimal path. With this goal in mind, we introduce a family of smoothly-varying curves which is general enough to incorporate both core widening and transition offset strategies.

11.4.3   Polynomial paths

We have seen that matching the fundamental-mode fields between adjoining waveguide segments of different curvatures, by offsetting the waveguide axes, reduces transistion loss. Another way of achieving the same result is to match the curvature of both segments, as well as the position and direction of their axes. If we impose all these conditions on a waveguide path connecting two positions in the x–z-plane, then a total of eight conditions would be required to specify the path, comprising two sets of x-, z-values for the positions of the ends of the path, together with two sets of slope (i.e. direction) and curvature values. One way to satisfy these conditions would be to represent x and z parametrically, i.e. set $x = x(t)$, $z = z(t)$ where t is a dummy variable. The waveguide path could then be represented by cubic, or third order, polynomials for both $x(t)$ and $z(t)$ in order to satisfy the eight end conditions. However, it is algebraically simpler to use a fifth order polynomial for both x and z to describe the path together with redundant end conditions as shown below.

We also need an optimization procedure to choose the lowest-loss path within the family, which can be achieved by using the following parametric representation in t:

$$r(t) = \begin{cases} x(t) = \displaystyle\sum_{n=0}^{5} a_n t^n \\[4mm] z(t) = \displaystyle\sum_{n=0}^{5} b_n t^n \end{cases} \tag{11.9}$$

The 12 polynomial coefficients a_i and b_i are then determined by the values of position, tangential direction and curvature of the core axis at each end of the path. If we define the arc length via the expression

$$s(t) = \int_0^t \left| \frac{d\mathbf{r}(t)}{dt} \right| dt \tag{11.10}$$

where the position vector $\mathbf{r} = (x, z)$, then the slope condition assumes the simple form

$$\dot{x}(s) = \sin \theta \tag{11.11a}$$
$$\dot{z}(s) = \cos \theta \tag{11.11b}$$

where the 'dot' denotes derivative with respect to s, while θ is the orientation of the tangential direction relative to the z-axis. Since the arc length s is the independent variable in (11.11), these expressions are correct for any value of s along the path. However, in terms of the variable t, a factor dt/ds appears on the right hand side of (11.11a) and (b). Setting $dt/ds = 1$ at both the beginning and end of the path has the effect of restricting the family of curves, without affecting the value of the slope angle θ. Hence, for our restricted family, we can write

$$\dot{x}(t_0) = \sin \theta \tag{11.12a}$$
$$\dot{z}(t_0) = \cos \theta \tag{11.12b}$$

where the 'dot' now denotes derivative with respect to t, and t_0 is the value of t at either the beginning or end of the path. Similarly, the curvature κ now satisfies

$$\ddot{x}(t_0) = \kappa \cos \theta \tag{11.13a}$$
$$\ddot{z}(t_0) = -\kappa \sin \theta \tag{11.13b}$$

where κ is the curvature required at t_0. Likewise, the curvature at any position along the path is given by

$$\kappa(t) = \frac{\ddot{x}(t)\dot{z}(t) - \ddot{z}(t)\dot{x}(t)}{[\dot{x}^2(t) + \dot{z}^2(t)]^{3/2}} \tag{11.14}$$

The coefficients appearing in (11.9) are readily evaluated from the solution of two linear systems of six coupled equations, in terms of the 12 end conditions.

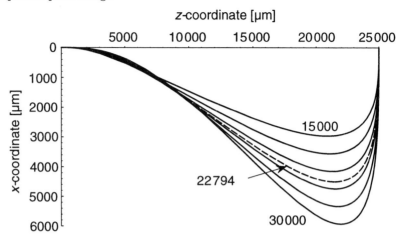

Figure 11.3 The group of polynomial paths satisfying the end-path conditions given in the text. The dotted curve represents the path that minimizes the total integrated bend loss of (11.17a).

If we assume that $t = 0$ corresponds to the beginning of the polynomial path, then the value of t corresponding to the end of the path, denoted by L as it is directly related to the length of the path, can be chosen arbitrarily. Accordingly, if the set of six linear equations for the coefficients of $x(t)$ in (11.9) is solved, the expression for the a_i can be expressed in matrix form as

$$
\begin{bmatrix} a_0 \\ a_1 \\ a_2 \\ a_3 \\ a_4 \\ a_5 \end{bmatrix} = \begin{bmatrix} 1 & 0 & 0 & 0 & 0 & 0 \\ 0 & 1 & 0 & 0 & 0 & 0 \\ 0 & 0 & 1/2 & 0 & 0 & 0 \\ -\frac{10}{L^3} & -\frac{6}{L^2} & -\frac{3}{2L} & \frac{10}{L^3} & -\frac{4}{L^2} & \frac{1}{2L} \\ \frac{15}{L^4} & \frac{8}{L^3} & \frac{3}{2L^2} & -\frac{15}{L^4} & \frac{7}{L^3} & -\frac{1}{L^2} \\ -\frac{6}{L^5} & -\frac{3}{L^4} & \frac{1}{2L^3} & \frac{6}{L^5} & -\frac{3}{L^4} & \frac{1}{2L^3} \end{bmatrix} \begin{bmatrix} x_i \\ \dot{x}_i \\ \ddot{x}_i \\ x_f \\ \dot{x}_f \\ \ddot{x}_f \end{bmatrix} \tag{11.15}
$$

where the subscripts i and f refer to the initial and final values at the beginning and end of the path, respectively. An identical expression applies to the coefficients b_i of the $z(t)$ polynomial where the b_is replace the a_is and z replaces x.

Figure 11.3 illustrates a typical family of polynomial paths used for the design of a progressive bend. The end conditions were zero curvature and zero slope at the beginning of the path ($\theta = 0$ and $\kappa = 0$) and zero curvature and 90° slope at the end of the path ($\theta = \pi/2$ and $\kappa = 0$). As L increases from 15 000 to 30 000, successive polynomial paths become longer.

11.4.4   Bend loss minimization

Once the end conditions for the set of polynomial paths are fixed, as in the example given in Figure 11.3, we need a strategy for determining which of the family of paths has the lowest overall bend loss, corresponding to $L = L_{\text{opt}}$, say. The power remaining in the fundamental mode at distance s along the path, $P(s)$, can be expressed in terms of the initial power $P(0)$ as

$$P(s) = P(0)\exp\left(-\int_0^s \gamma(s)\mathrm{d}s\right) \tag{11.16}$$

where γ is the local power attenuation coefficient distance s along the path, and the integral in the exponent represents the accumulated bend loss over distance s. We assume that the curvature along each path varies sufficiently slowly that the expression for γ is well approximated by the pure bend loss expression given by (11.5), where the constant bend radius R_c is replaced by the local bend radius $R_c(s)$.

Determination of the lowest-loss path is then equivalent to minimizing the integrated attenuation coefficient $I(L)$, where

$$I(L) = \int_0^L \gamma(t)\left|\frac{\mathrm{d}\mathbf{r}}{\mathrm{d}t}\right|\mathrm{d}t \tag{11.17a}$$

by requiring that

$$\frac{\mathrm{d}I(L)}{\mathrm{d}L} = 0 \tag{11.17b}$$

This procedure can be carried out efficiently using standard numerical procedures to first calculate (11.17a) and then solve (11.17b) for L_{opt}. When the minimization is carried out for the polynomial family in (11.3), the dashed path minimizes $I(L)$ and corresponds to $L_{\text{opt}} = 22794$. If one were to take into account the two-dimensional aspect of a particular type of waveguide, the attenuation coefficient could be chosen as in Marcuse (1976).

Using the calculus of variations, one could, in principle, solve for the path $\mathbf{r}(t)$ that minimizes the functional (11.17). Although some approximate solutions have been proposed (Mustieles $et\ al.$, 1993) for particular configurations, the end conditions are not easily matched for an arbitrary path, and discontinuous offsets are still required, because transition loss is not taken into account in this approach. If the attenuation coefficient γ of (11.5) is replaced by the simpler, but less accurate form κ^2, the minimization problem (11.17) becomes analytical and leads to paths described by incomplete elliptic functions (Ladouceur, 1991).

The polynomial-path method described here circumvents transition loss, since curvature is continuous everywhere along the path, and does not require any core widening for its implementation. However, if core widening is also incorporated, a much more flexible technique is produced; this is the basis of the following section.

11.5   LOSS MINIMIZATION AND CORE WIDENING

We showed in Section 11.4.1 that transition loss between waveguides of different curvature is substantially reduced by offsetting one of the cores. If, instead of an abrupt change in curvature, we consider a waveguide path with continuously varying curvature, then the smooth path can be regarded as comprising an infinite number of infinitesimally short waveguides, each with a slightly different radius of curvature.

Figure 11.4 illustrates this description for two infinitesimal adjacent segments of the path. Without loss of generality, the first segment in (a) has zero curvature and the second an infinitesimal curvature $d\kappa$. The maxima of the two modal fields will coincide if the second waveguide core is offset a distance dx_d in (b), given by (11.6) with $R_c = 1/d\kappa$. Introducing an infinitesimal offset between two infinitesimal segments to align the modal fields, as in Figure 11.4b, amounts to redefining a different path with no offset in the limit where the segment length $ds \to 0$. Thus we need an alternative strategy.

This can be achieved by widening the core of the second segment by an infinitesimal amount equal to twice the offset, i.e. $2dx_d$, as illustrated in Figure 11.4c. This widening displaces the axis of the core by just the offset dx_d towards the centre of curvature, so that the field of the second segment is then approximately aligned with that of the first segment.

As well as aligning the modal field between successive segments, widening the waveguide core also reduces the pure bending attenuation coefficient γ in (11.5), so that bending loss decreases. Secondly, the fundamental mode field in the cross-section of the bent waveguide is slightly compressed, as well as being offset (Garth, 1987; Marcuse, 1973). Hence widening of the core stretches the field to compensate for this effect, and also improves the mode-matching between adjacent segments.

This analysis leads to a very simple numerical scheme. For a givenwaveguide, the required widening, $2dx_d$, for a particular path is calculated from (11.6), and is readily shown to be proportional to the local curvature

$$2dx_d = C\kappa, \quad C = \frac{2V^2}{\Delta}\frac{s_x^4}{\rho^2} \tag{11.18}$$

where the mode spot size s_x is determined from (11.8).

11.5.1   Core widening and adiabaticity

It is necessary to ensure that the core widening is not too rapid, as a severe uptaper could lead to non-adiabatic behaviour, whereby the fundamental mode would couple power directly with the radiation field, as well as losing power through bending. It has been shown (Love, 1989) that for a taper to be

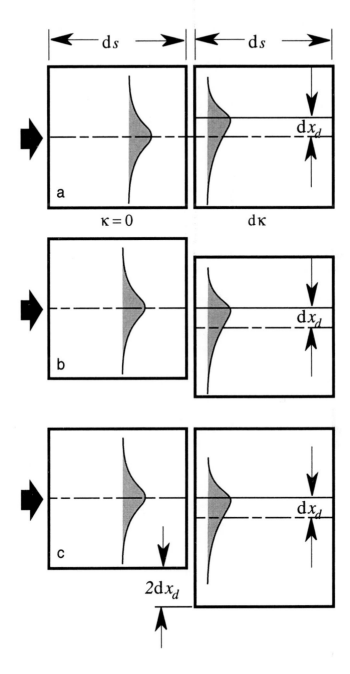

Figure 11.4 Field matching strategy: (a) infinitesimal mode offset dx_d due to an infinitesimal change of curvature $d\kappa$; (b) compensation by an equal offset of the waveguide; (c) compensation by widening the core by $2dx_d$.

approximately adiabatic, the local taper angle Ω should satisfy

$$\Omega(s) < \frac{k\rho(s)(n_{\text{eff}} - n_{\text{cl}})}{2\pi} \tag{11.19}$$

where the taper angle $\Omega(s)$, assumed small, is approximately equal to $d\rho/ds$, and the remaining parameters are defined elsewhere. It also follows from (11.18) that $d\rho/ds = Cd\kappa/ds$, so the value of the constant C satisfies

$$C < \frac{k\rho(s)(n_{\text{eff}} - n_{\text{cl}})}{2\pi} \left(\frac{d\kappa}{ds}\right)^{-1} = \frac{\sqrt{2\Delta}}{4\pi} \frac{W^2}{V} \left(\frac{d\kappa}{ds}\right)^{-1} \tag{11.20}$$

This sets a limit on the curvature that can be compensated for, and thus sets arestriction on the possible paths, as the maximum value of C depends on the termdκ/ds, which is path dependent.

11.5.2   Combined loss optimization and core widening

When calculating the optimal curve within the polynomial family, we used a minimization procedure on the functional $I(L)$ given by (11.17a). The parameters involved in $I(L)$ are calculated by solving the eigenvalue equation (11.3) together with (11.1) to (11.4) for the fundamental mode parameters, including the effective index n_{eff}.

To incorporate the widening strategy of the previous section, we calculate $I(L)$ using the values of effective index and related parameters at each point along the waveguide path calculated from (11.3). This approach is computationally straightforward, but can be simplified by using a Taylor series expansion of the normalized eigenvalue, W, from (11.3) about its unperturbed value W_0

$$W(t) = W_0 + (\rho(t) - \rho_0)\left(\frac{dW}{d\rho}\right) \tag{11.21a}$$

where ρ_0 is the unperturbed half-width and (Snyder and Love, 1983)

$$\frac{dW}{d\rho} = \frac{1}{\rho}\frac{V^2 + W}{1 + W} \tag{11.21b}$$

We then have a general procedure for generating loss-optimized paths between arbitrary points on an optical chip. This procedure can be summarized as follows.

(i) Define the position, tangent and curvature at each end of the path.
(ii) Calculate the widening coefficient C using (11.18).
(iii) Calculate the loss functional $I(L)$ using (11.17).
(iv) Find the minimum of $I(L)$.

This procedure is readily automated for arbitrary end conditions (LÉTI, 1994) and has been implemented in a mask-generating software (Cadence,

Table 11.1 Optimization of the concatenated path for the 90° bend in Figure 11.5. Linear dimensions are in microns and losses in dB. The total loss is the sum of the pure bend loss (PB-loss) and the transition loss (TL-loss) at the two interfaces.

Widening (μm)	Offset (μm)	TL-loss (dB)	PB-loss (dB)	Total loss (dB)	BPM-loss (dB)
0.0	0.00	0.094	0.507	0.694	0.306
0.0	0.55	0.047	0.507	0.600	0.298
0.5	0.55	0.033	0.297	0.362	—
1.0	0.60	0.266	0.183	0.236	—
1.5	0.65	0.025	0.119	0.168	—
2.0	0.65	0.027	0.080	0.134	—
2.5	0.70	0.032	0.055	0.119	0.109
3.0	0.80	0.039	0.039	0.117	0.112
3.5	0.85	0.048	0.030	0.127	—

1993). In order to quantify the effectiveness of the procedure, we compare the loss of the optimum polynomial widened path with an optimized concatenation of straight and circular segments in the following section.

11.6   OPTIMAL LOW-LOSS 90° BEND

A path of particular interest in the design of planar circuitry is the 90° bend, as illustrated in Figure 11.5. It is composed of two straight segments forming the input and output ports and a curved path turning through 90°. Figure 11.5 a shows the situation where an arc of acircle of constant curvature has been used to form the bend, while Figure 11.5b shows the optimal polynomial-based path. The overall loss for the former is due to the transition losses at the two straight–curved interfaces and the pure bend loss along the arc. As explained in Section 11.4.1, we have optimized the overall loss by introducing offsets at the splices A and B and by widening the circular arc from the nominal width of 6.5 μm of the straight waveguide. Table 11.1 presents the results from this optimization procedure.

Table 11.1 shows the optimal low-loss bend with a widening of 3 μm and offsets of 0.8 μm. Note the substantial reduction in total loss from 0.694 dB to 0.117 dB, compared with the no-offset, no-core widening design. The loss figures have also been compared with the corresponding results of a BPM simulation, shown in the last column of Table 11.1. The lower losses predicted by the BPM

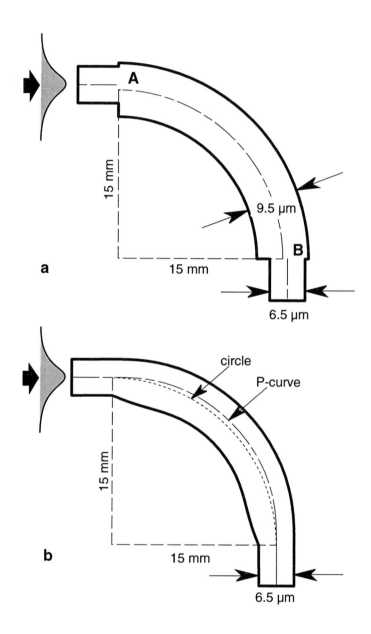

Figure 11.5 Two 90° bends: (a) the optimized concatenated bend with offsets and core widening; (b) the optimized polynomial path with variable core width.

Table 11.2 Optimization of the continuous path for the case of the 90° bend illustrated in Figure 11.5b. The widening coefficient was set to zero in the first row, and was automatically produced according to (11.18) in the second row

Free parameter L	Widening coefficient C $[\mu m^2]$	BPM loss [dB]
21 726	0	1.18
21 793	23 837	0.10

are due to recoupling of power from the radiation field to the fundamental mode along the bend, as this is not accounted for in the splicing approach.

The results for the corresponding optimal polynomial path are presented in Table 11.2. The first and second rows give the optimal path parameter L without and with widening, respectively. These paths were obtained using an automated numerical routine, while the corresponding total loss obtained from the BPM is lower than the figure for the concatenated arrangement presented in Table 11.1. Note that the optimal path obtained without widening has a loss greater than that of the simple circular path without widening or offset. This shows that the mode mismatch, due to curvature, along the continuous path is a more important effect. It is more than compensated for by the widening procedure discussed in Section 11.5.

11.6.1 Modal field comparison

Further insight into propagation along the two paths in Figure 11.5 can be obtained by examining Figure 11.6, where the output fundamental-mode intensity distribution over the waveguide cross-section is superposed on the input intensity distribution for both the concatenated and continuous paths. Noting that pure bend loss is comparable in both cases (around 0.1 dB), the output field in the concatenated case is quite distorted compared with that for the continuous case, showing that (i) there is slightly more pure bending loss within the continuous case than in the concatenated case, and (ii) that transition loss accounts for most of the loss in the concatenated case. A further consequence of (ii) is that the light radiating from the bend in the concatenated case, as essentially forward-directed radiation from the mismatch, will beat with the guided power in the fundamental mode over possibly long distances, whereas this is not the case for the continuous path.

BPM calculations have also shown that the loss along the optimized continuous path can be slightly reduced by trial and error, by varying the values of L and C around the actual values obtained by our procedure. In most cases,

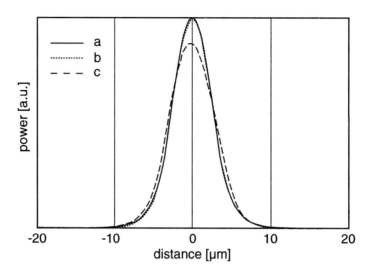

Figure 11.6 Modal intensity distribution comparisons: (a) and (b) show, respectively, the input and output fields of the continuous bend of Figure 11.5(b), while (c) shows the corresponding fields of the concatenated bend of Figure 11.5a.

this further optimization is not normally necessary but can always be carried out if bending loss along the path is considered critical.

11.7   APPLICATION TO PRACTICAL OPTICAL PATH DESIGN

The procedure outlined above has more general application to the design of other types of low-loss curved paths, such as S-bends. This optimization has, in all cases that have been tested, produced loss estimates that are as low as or lower than equivalent concatenated paths. This method greatly facilitates the design of complex structures, such as power splitters, using cascaded Y-junctions, free-space dividers (Day *et al.*, 1992), star couplers (Dragone *et al.*, 1989), and mode filters based on S-bends etc.

Another useful outcome in using continuous paths is related to fabrication-errors and tolerances. As the paths are continuous, the effect of over- or under-dimensioning, due to imperfect control of the etching of BCWs, should be lower than that for the concatenated paths. Similarly, an imperfect control of the refractive index value has a direct effect on the offset, while the effect for continuous guides, being approximately adiabatic, should be reduced.

REFERENCES

Cadence (1993) *Cadence version 4.2.1.* San-Jose, California: Cadence Design System, Inc.

Day, S., Bellerby, R., Connell, G. and Grant, M. (1992) Silicon based fibre pigtailed 1×16 power splitter. *Electronics Letters*, **28**, 920–922.

Dragone, G., Henry, C. H., Kaminow, I. P. and Kistler, R. C. (1989) Efficient multichannel integrated optics star coupler on silicon. *IEEE Photonics Technology Letters*, **1**, 241–243.

Garth, S. J. (1987) Modes on a bent optical waveguide. *IEE Optoelectronics Part-J*, **134**, 221–229.

Ladouceur, F. (1991). *Buried Channel waveguides and devices.* Ph.D. thesis, Australian National University.

Ladouceur, F. and Labeye, P. (1995) A new general approach to optical waveguide path design. *IEEE Journal of Lightwave Technology*, **LT–3**, 481–492.

LÉTI (1994). The algorithm summarized in this section is subject to a French patent to be extended in the near future by LÉTI, Grenoble, France.

Love, J. D. (1989) Application of low-loss criterion to optical waveguides and devices. *IEE Optoelectronics Part-J*, **136**, 225–228.

Marcuse, D. (1973) *Light transmission optics.* New York: Van Nostrand Reinhold Company.

Marcuse, D. (1976) Field deformation and loss caused by curvature of optical fibers. *Journal of the Optical Society of America*, **66**, 311–320.

Mustieles, F. J., Ballesteros, E. and Baquero, P. (1993) Theoretical S-bend profile for optimization of optical waveguide radiation losses. *IEEE Photonics Technology Letters*, **5**, 551–553.

Neumann, E. G. (1982) Curved dielectric optical waveguides with reduced transition losses. *IEEE Proceedings Part-H*, **129**, 278.

Pennings, E. C. M. (1990). *Bends in optical ridge waveguides (modeling and experiments).* Ph.D. thesis, Technische Universiteit Delft.

Snyder, A. W. and Love, J. D. (1983) *Optical waveguide theory.* London: Chapman & Hall.

12
Single-mode planar couplers

The optical fibre coupler is an established device which has been refined continuously over the last 20 years, so that a wide variety of commercial quality, very low-loss couplers is now available for a diversity of optical splitting and combining processes. The majority of these couplers is fabricated using the fused-taper technique. However, there are several goals which the tapered fibre coupler has not been able to achieve.

Firstly, there is an inherent lower limit to the overall dimensions of a single tapered coupler. Although the central coupling region of a well-fused tapered fibre coupler is only of the order of a few millimetres long, when combined with the necessarily relatively long length of the adiabatically tapered input and output fibres and packaging, it leads to an overall device length of several centimetres. Secondly, in a concatenation of fibre couplers, the two output leads from the first coupler need to be spliced to the input ports of the second and third couplers, and so on. Further, there is a lower limit on the bending radius of fibres, of the order of a few centimetres, below which the joining leads cannot be bent in order to avoid significant radiation loss, and this places a lower limit on the overall lateral dimensions of the concatenation layout.

Both of these limitations can be overcome through the use of planar, or buried channel couplers. The overall length of a planar coupler is much shorter than that of the corresponding fibre coupler, and concatenations of planar couplers occupy a very much smaller area, typically of the order of a few cm^2. Furthermore, the complete concatenated of coupler system can be fabricated as a single unit.

Here, we concentrate on single-mode couplers, which, following the usual convention, comprise two single-mode waveguides (when in isolation of each other). These couplers are, in general, two-moded waveguides in spite of their nomenclature. Few- and multi-mode couplers supporting more than two modes can also be analysed using the techniques presented in this chapter.

12.1 COUPLER FABRICATION

Although fibre couplers can be fabricated using polishing techniques, the more common approach uses two fibres fused into a single tapered element, using

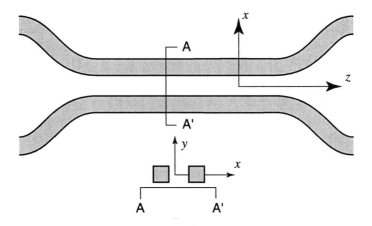

Figure 12.1 Schematic of a planar waveguide coupler with curved input/output arms and mid-waist square-core cross-section.

a combination of heating and drawing. One of the advantages of the latter technique is the ability to monitor optically the coupling behaviour during fabrication, and hence terminate the process when the requisite coupling behaviour is achieved. In other words, the production of fused-taper couplers with specific properties relies essentially on fabrication expertise and requires little theoretical input.

Planar couplers differ from their fibre counterparts in two significant respects. Firstly, fused taper couplers rely on interference between cladding modes. i.e. modes whose effective index values are below the cladding index, for their operation, whereas planar couplers are essentially untapered devices with an infinite cladding, and rely on interference between guided core modes, i.e. modes whose effective-index values lie between the maximum core index and the cladding index. In the former case, the coupling length between cores is reduced to millimetres through tapering, whereas in the latter case, a comparable coupling length can be achieved by designing the cores to be arbitrarily close together.

Secondly, planar device performance cannot be monitored during the fabrication process. Thus, the coupler must be completely designed theoretically beforehand. This is not a straightforward exercise, as the coupling length is dependent on all the source and coupler parameters. Consider the schematic of the planar waveguide coupler design shown in Figure 12.1, where the cores are of constant cross-section and the curved regions lead into the straight central coupling region. As well as prescribing values for source wavelength, core size and index, and core separation, the design must take account of the contribution to coupling in the curved input and output arms, as well as in the central coupling region, and must also ensure that the curvature in the arms is not too large to incur significant bend loss. These considerations are analysed in more detail below.

12.2   PHYSICAL PHENOMENON OF RESONANCE

Coupling between the fundamental mode of one single-mode optical waveguide and the fundamental mode of a second waveguide with an identical propagation constant is based on the general physical principle of resonance. This phenomenon occurs when the frequencies in two separate systems become equal. One well-known example of resonance is the wine glass and the opera singer. The wine glass has a set of frequencies at which it will ring, as can be demonstrated by rubbing the rim with a wet finger. If the opera singer hits a note at the same frequency as one of the glass, energy is transferred to the wine glass through sound waves. The material of the glass is relatively rigid, so that it can only accommodate limited energy, and the amplitude of oscillation is thus small. Once this limit is exceeded, the glass shatters.

The corresponding electromagnetic resonance phenomenon accounts for the transfer of light power propagating along one optical waveguide to a parallel second waveguide and vice versa. If we consider single-mode waveguides, which propagate only the fundamental mode, it is intuitive that we will have a resonance situation if the two waveguides are identical, since the fundamental mode of each fibre will have the same propagation constant. Thus the simplest electromagnetic resonance configuration, or coupler, would consist of two identical, single-mode BCWs laid parallel to one another, as shown in Figure 12.2.

For convenience, we refer to the coupler in Figure 12.2 as a *uniform coupler*, because its cross-section is invariant along its length, and the coupler in Figure 12.1 as a nonuniform coupler because of its bent input/output arms.

12.3   MODEL FOR THE UNIFORM SYMMETRIC COUPLER

Figure 12.2 shows a two-dimensional plan of the model for describing the symmetric coupler. We assume a step profile, so that both cores have refractive index n_{co} and square cross-section with half-width ρ, surrounded by a uniform, unbounded cladding of index n_{cl}. The separation of the two core axes is d. The positions where optical power enters and leaves the coupler are referred to as 'ports' and are numbered 1–4 clockwise from the top left-hand port, which is normally taken to be the input port.

Light enters the coupler along the core of one waveguide as the fundamental mode through port 1. Because of the close proximity of the identical second

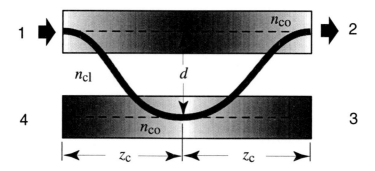

Figure 12.2 Model of the uniform single-mode waveguide coupler comprising two parallel BCWs.

core, the cladding field of the mode overlaps the core of the second waveguide, and a resonance is set up. Light couples from the first core across to the second core as it propagates along the coupler, denoted by the thick curve, and the power in the mode of the first waveguide is coupled to the field of the mode of the second waveguide. All of the light couples across to the second core in a distance z_c, which is the *coupling length*, measured along the coupler z-axis.

When all of the light has coupled to the second waveguide, the coupling argument can be repeated. The close proximity of the identical first core to the second core means that the field of the fundamental mode of the second waveguide overlaps the core of the first waveguide, and resonance occurs between the two. Thus the whole process is reversed and all the light couples back to the first core in distance z_c. In other words, light has coupled to the second core and back in distance $2z_c$, referred to as the *beat length*, z_b. This behaviour is then repeated periodically along the length of the coupler.

It follows from the oscillation of light between the two cores of the coupler in Figure 12.2 that the distribution of light between the two output ports 2 and 3 depends on whether the length of the coupler, z_L, is an integer multiple of the coupling length or not. If z_L is an odd integer multiple of z_c, then all of the power entering port 1 of the first fibre leaves the coupler in the second fibre through port 3. Similarly, if z_L is an even integer multiple of z_c, all the power exits through port 2 in the first fibre, as if no coupling had occurred. If z_L is not an integer multiple of z_c, then some light leaves the coupler through port 3 and the rest through port 2.

The coupling, or beat length, determines the overall length of a coupler, for specific applications, so it is therefore necessary to quantify z_c. There are two ways in which this analysis can be carried out, using either coupling between the fundamental mode of each core, i.e. coupled-mode theory, discussed in Section 12.4, or a superposition of the modes of the composite two-core coupler waveguide, i.e. supermode theory, discussed in Section 12.5. The coupling length depends on all the physical parameters of the coupler, namely core size, relative index difference, core separation, as well as the wavelength of the source.

12.3.1 Mechanical Analogue

The coupling phenomenon can be demonstrated through a simple mechanical analogue. The wavelike propagation of the fundamental mode along a waveguide is the electromagnetic analogue of an oscillating simple pendulum, as the periodic swinging of the pendulum in time corresponds to the sinusoidal variation of the mode with distance along the waveguide through the longitudinal dependence $\exp(i\beta z)$. Accordingly, the mechanical analogue of the coupler consists of two coupled identical simple pendula attached to a third horizontal string which provides the mechanical coupling between the two pendula, just as the cladding allows electromagnetic coupling between the fields of the fundamental modes of the two waveguides. The separation of the strings corresponds to the separation of the two cores. If the first pendulum is set off, with the second pendulum being stationary, then the horizontal string slowly transfers all of the mechanical energy to the second pendulum, the time taken for this transfer being the analogue of the coupling length z_c.

12.4 COUPLED MODE SOLUTION

The analytical expressions for coupling between two parallel, uniform waveguides, of otherwise arbitrary cross-section and core refractive-index profile, are well known, and can be derived from coupled mode theory. In this approach, the effect of the second core of the coupler is to modify the amplitude of the fundamental mode propagating along the first core, and, conversely, the presence of the first core modifies the amplitude of the fundamental mode propagating in the second core. This mutual interaction between the two fundamental-mode amplitudes along the coupler can be described by a pair of coupled equations. Assuming unit power entering port 1, at wavelength λ, in Figure 12.2, the solutions for the powers P_2 and P_3 leaving ports 2 and 3, respectively are readily shown to be given by (Snyder and Love, 1983, Section 18–13)

$$P_2 = \cos^2(Cz_L), \quad P_3 = \sin^2(Cz_L) \tag{12.1}$$

where z_L is the length of the coupler and $C(\lambda)$ is the wavelength-dependent coupling coefficient. This solution conserves power along the coupler, as $P_2 +$

$P_3 = 1$, independent of z_L. Assuming a step profile, with core and cladding indices n_{co} and n_{cl}, respectively, the coupling coefficient is defined by

$$C = k(n_{co} - n_{cl})\frac{\int_{A_{CO2}} \psi_1\psi_2 dA}{\int_{A_\infty} \psi_1^2 dA} \tag{12.2}$$

where $k = 2\pi/\lambda$ is the free-space wavenumber, λ the source wavelength, A_{CO2} the core cross-section of the second waveguide and A_∞ the infinite cross-section of the coupler. The functions ψ_1 and ψ_2 are the solutions of the scalar wave equation for the fundamental mode fields of waveguides 1 and 2, respectively, in isolation from one another. The normalization is chosen so that, relative to coordinates based on the respective core axes, they would have identical functional forms. In (12.2), however, they are referred to the common x–y-axes in Figure 12.1, in which case their functional forms will be different.

The beat and coupling lengths are expressible in terms of the coupling coefficient by

$$z_b = 2z_c = \frac{\pi}{C} \tag{12.3}$$

Qualitatively, the coupling coefficient decreases with increasing core separation d, the coefficient varying as $\exp(-Wd/\rho)$ for large separations, where W is the cladding mode parameter. Consequently, the coupling and beat lengths both increase rapidly with increasing core separation, demonstrating the need for close proximity of the two cores to achieve practically short couplers. An analytical expression is available for C for two step-profile circular cores, which is accurate for all values of separation (Snyder and Love, 1983, Section 18-14), but no equivalent expression is available for BCW cores. Accordingly, the latter must be determined numerically using, e.g. either the FDM or MFDM.

12.5 SUPERMODE SOLUTION

The alternative approach to determining propagation is based on the modes of the composite two-core waveguide defined by the coupler. To avoid confusion with the modes of the single-core waveguide, the modes of the two-core waveguide are sometimes referred to as 'supermodes', as discussed in Section 3.9. Since each core in isolation from the other is single-moded, the two-core waveguide is usually two-moded for practical devices. This aspect is discussed further below.

In the supermode approach, the coupling of power between the two cores is described in terms of the superposition of the two guided supermodes. The symmetry of the coupler about its mid-plane in Figure 12.2, requires that one supermode, the fundamental supermode, has a field symmetric about this plane, while the second supermode must have an antisymmetric field. Accordingly, they have the respective representations

$$a_+\psi_+(x,y)e^{i\beta_+z}, \quad a_-\psi_-(x,y)e^{i\beta_-z} \tag{12.4}$$

where a_+ and a_- are constants, the supermode amplitudes, ψ_+ and ψ_- are even and odd solutions, respectively, of the scalar wave equation for the two-core waveguide, and β_+ and β_- are the corresponding propagation constants, solutions of the eigenvalue equation for the even and odd supermodes, as determined from the scalar wave equation. Thus the field ψ at position z along the coupler is expressible as

$$\psi(x,y,z) = a_+\psi_+(x,y)e^{i\beta_+z} + a_-\psi_-(x,y)e^{i\beta_-z} \tag{12.5}$$

where a_+ and a_- are determined from the excitation condition at the beginning of the coupler, $z = 0$. It follows immediately from this expression that the beat and coupling lengths are given by

$$z_b = 2z_c = \frac{2\pi}{\beta_+ - \beta_-} \tag{12.6}$$

and, consequently, by comparison with (12.3), that

$$\beta_+ - \beta_- = 2C \tag{12.7}$$

This links the coupled-mode and supermode solutions, so that the coupling coefficient is half the difference between the supermode propagation constants.

When the two cores are well separated, the supermode fields ψ_+ and ψ_- are well approximated by the sum and difference of the fundamental-mode fields of the respective cores in isolation from one another, i.e.

$$\psi_+ = \psi_1 + \psi_2, \quad \psi_- = \psi_1 - \psi_2 \tag{12.8}$$

provided ψ_1 and ψ_2 have the same normalization. Using this relationship, together with (12.5), it is straightforward to show that if port 1 in Figure 12.2 is excited by the fundamental mode of core 1 with unit power, the resulting expressions for the power in each core at distance z along the coupler are identical to those in (12.1).

Using the supermode approach to determine the propagation characteristics of symmetric BCW couplers with square cores, of side 2ρ, requires a numerical solution of the scalar wave equation to determine both the fields and propagation constants. This can be readily carried out using either the FDM or MFDM.

12.5.1   Odd-supermode cutoff

In the supermode analysis of propagation, the fundamental supermode propagates for any value of source wavelength, because of the simple matched-cladding refractive-index profile, i.e. the index throughout both cores is higher than the surrounding cladding index. However, by analogy with the second

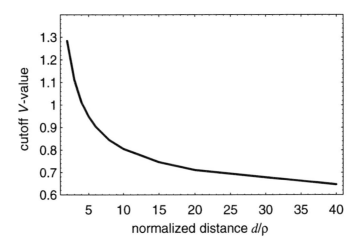

Figure 12.3 Plot of the odd supermode cutoff V-value against normalized core separation.

mode of a single-core waveguide, the second supermode of the twin-core coupler will always have a finite cutoff wavelength. This cutoff wavelength can be a constraint in the design of couplers with specific coupling or splitting properties, e.g. the broadband couplers to be described in Section 12.9.

The cutoff wavelength of the odd supermode can be determined analytically for circular cores (Love and Ankiewicz, 1985), but no corresponding solution is available for BCW cores, and hence numerical techniques must be employed. Using the MFDM, the cutoff V-value, V_c, of the odd supermode is plotted against normalized separation d/ρ in Figure 12.3 (Hewlett *et al.*, 1995). If λ_c denotes the cutoff wavelength, it is related to V_c by

$$V_c = \frac{2\pi}{\lambda_c}\rho(n_{co}^2 - n_{cl}^2)^{1/2} \tag{12.9}$$

in terms of the core half-size ρ, and the core and cladding indices n_{co} and n_{cl}, respectively.

An accurate, analytical approximation to the numerical value of V_c can be generated by assuming that the core area of the BCW is equal to that of the fibre, i.e. $\rho_f = 2\rho/\sqrt{\pi}$, where ρ_f is the fibre core radius. If this expression for the core radius is substituted into the approximate expression for the fibre cutoff value of V_c (Love and Ankiewicz, 1985), we obtain

$$V_c = \left[\frac{2\pi}{1 + 4\ln\left(\frac{\sqrt{\pi}}{2}\frac{d}{\rho}\right)}\right]^{1/2} \tag{12.10}$$

When compared with the numerical values for V_c in Figure 12.3, the approximate values given by 12.10 are in error by less than 1% for $d/\rho > 4$.

12.6   ASYMMETRIC COUPLERS

An asymmetric BCW coupler has two cores which are uniform but slightly different. The difference can be in the cross-sectional geometry, refractive-index profile, or both, but must be slight to ensure that the propagation constants β_1 and β_2 of the fundamental modes of cores 1 and 2, respectively, are still sufficiently similar for significant coupling of power to occur. In the symmetric case, $\beta_1 = \beta_2$ and 100% power swapping between cores occurs, but as the difference between β_1 and β_2 increases, the fraction of power coupled decreases very rapidly. Assuming unit power entering port 1 of the asymmetric coupler, then either the coupled-mode or supermode approaches lead to (Snyder and Love, 1983, Section 18-20)

$$P_2 = 1 - F^2 \sin^2\left(\frac{C}{F}z_L\right), \quad P_3 = F^2 \sin^2\left(\frac{C}{F}z_L\right) \qquad (12.11)$$

for the power leaving ports 2 and 3, respectively, where F^2 is the maximum fraction of input power coupled between the two cores, which is expressible as

$$F^2 = \frac{1}{1 + \left(\frac{\beta_1 - \beta_2}{2C}\right)^2} \qquad (12.12)$$

where the coupling coefficient, C, is expressible in terms of the mode and supermode propagation constants as

$$C(z) = \left\{[\beta_+(z) - \beta_-(z)]^2 - (\beta_1 - \beta_2)^2\right\}^{1/2}/2 \qquad (12.13)$$

and satisfies $C/F = (\beta_+ - \beta_-)/2$.

12.7   NONUNIFORM COUPLERS

Propagation through the nonuniform single-mode planar coupler with curved input and output cores and uniform central region, as illustrated in Figure 12.4, can be described analytically using either coupled modes or local supermodes. Here we adopt the latter approach to provide more physical insight, since the supermodes are the natural resonances of the two-core device. The simplest design assumes constant radius bends, as shown in Figure 12.4, with bend radius R_c sufficiently large to avoid significant bend loss. Given values for the core parameters and source wavelength, a suitable value for R_c is readily determined from Chapter 10. It is convenient to consider the input, central and output

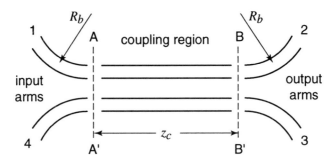

Figure 12.4 Division of the nonuniform coupler into three regions.

regions of the coupler separately. These are bounded by the dashed lines AA' and BB'.

12.7.1   Input region

Assuming unit power input entering port 1 in the input region of Figure 12.4, there will be negligible coupling between the two arms until they are relatively close together, because the coupling coefficient $C(d)$ decreases exponentially with increasing separation $d(z)$ between the core axes. The interaction along the curved sections can be calculated using either coupled local-mode equations, or, equivalently, the superposition of the two local supermodes. Local supermodes were discussed in Section 3.9. As stated above, we choose the latter approach mainly because of the more physical interpretation available in using the modes of the composite two-core waveguide.

12.7.2   Local supermodes

Propagation along the curved BCW input arms in Figure 12.4, and along the corresponding output arms can be described by a superposition of the even and odd local supermodes, to account for the continuous variation in separation $d(z)$ between the two cores. This assumes that each local supermode is adiabatic and that its power, and therefore amplitude, is conserved as it propagates in the curved region. This condition is virtually satisfied as there is no coupling between the two local supermodes, due to the symmetry of the coupler, and the bend radius is assumed large enough for bend loss to be negligible.

The forms of the local supermodes, as discussed in Sections 3.8 and 3.9, are generalisations of (12.4) (Snyder and Love, 1983, Section 19–1)

$$E_+ = b_+ \hat{\psi}_+(x, y, z) \exp\left(i \int \beta_+(z)dz\right) \qquad (12.14a)$$

$$E_- = b_- \hat{\psi}_-(x, y, z) \exp\left(i \int \beta_-(z)dz\right) \qquad (12.14b)$$

where b_+ and b_- are complex constants, so that $|b_+|$ and $|b_-|$ are the respective local mode amplitudes. Here $\beta_+(z)$ and $\beta_-(z)$ are the corresponding local propagation constants of the even and odd supermodes at position z along the axis of the input region, i.e. the solutions of the scalar eigenvalue equation for a uniform waveguide with the given cross-sectional geometry and profile at that position. The 'hat' indicates that the supermode fields are orthonormal, i.e.

$$\hat{\psi}_\pm(x, y, z) = \frac{\psi_\pm(x, y, z)}{N_\pm^{1/2}} \qquad (12.15a)$$

where N_+ and N_- are respective scalar normalizations defined by

$$N_\pm = \int_{A_\infty} \psi_\pm^2 dA \qquad (12.15b)$$

in terms of integration over the infinite cross-section of the coupler. The ψs and βs are calculated numerically at each position z along the input arms using the FDM or MFDM, from which the normalization and the phase integrals in (12.14) are also calculated numerically.

12.7.3   Excitation

The coupler is assumed to be excited by the fundamental mode through input port 1 only. To determine the amplitudes of the local supermodes in the input arms, the fields $\hat{\psi}_+$ and $\hat{\psi}_-$ of (12.15a) are simply related to the fundamental-mode fields ψ_1 and ψ_2 of the fundamental modes of core 1 and core 2 in isolation, respectively, when the two cores in Figure 12.4 are well separated (Snyder and Love, 1983, Section 18–12)

$$\psi_\pm = \psi_1 \pm \psi_2 \qquad (12.16)$$

Accordingly, since $N_+ = N_- = 2N$ in the limit of infinite core separation, where N is the common normalization for both ψ_1 and ψ_2, the amplitudes of the even and odd local supermodes must be equal, requiring that

$$|b_+| = |b_-| \qquad (12.17)$$

in (12.14), i.e. the modal amplitudes of the supermodes are equal.

12.7.4   Accumulated phase

As the two local supermodes propagate adiabatically along the curved input arms to the cross-section AA' in Figure 12.4, they accumulate a total relative phase difference ϕ_i given by

$$\phi_i = \int_{-\infty}^{AA'} [\beta_+(z) - \beta_-(z)]dz \tag{12.18}$$

where the lower limit of integration should be interpreted as meaning the value of z to the left of AA' below which there is negligible contribution to the integrand. As z decreases, $d(z)$ increases, and both β_+ and β_- rapidly approach the value of the fundamental-mode propagation constant β for each core. Note that ϕ_i is wavelength-dependent through the variation of the propagation constants with the source wavelength λ.

12.7.5   Central region

At AA', the local supermodes become the respective uniform supermodes of (12.4), and, since the modal amplitudes are constant in the respective regions, it follows from (12.15a) that

$$|a_+| = \frac{|b_+|}{N_+^{1/2}}, \quad |a_-| = \frac{|b_-|}{N_-^{1/2}} \tag{12.19a}$$

Furthermore, the accumulated relative phase difference ϕ_i between the bs at AA' introduces the same relative phase difference between the as, i.e.

$$a_- N_-^{1/2} = a_+ N_+^{1/2} e^{-i\phi_i} \tag{12.19b}$$

where N_+ and N_- are evaluated on AA'. Along the uniform central coupling region, there is a further increase in the relative phase difference between the two supermodes, ϕ_c, given by

$$\phi_c = (\beta_+ - \beta_-)z_L \tag{12.20}$$

where z_L is the length of the central coupling region between AA' and BB' in Figure 12.4. As with ϕ_i, this expression depends on source wavelength.

12.7.6   Output region

At BB', the uniform supermodes become the local supermodes of the bent output arms. If the latter have amplitude constants c_+ and c_-, respectively, at BB', then, by analogy with (12.19a), the modal amplitudes satisfy

$$|c_+| = |a_+| N_+^{1/2}, \quad |c_-| = |a_-| N_-^{1/2} \tag{12.21a}$$

and there is now an accumulated relative phase difference $\phi_i + \phi_c$, so that

$$c_- N_-^{1/2} = c_+ N_+^{1/2} e^{-i(\phi_i + \phi_c)} \tag{12.21b}$$

Finally, by analogy with (12.18), the total accumulated relative phase difference along the output arms, ϕ_o, is given by

$$\phi_o = \int_{BB'}^{\infty} [\beta_+(z) - \beta_-(z)] dz \tag{12.22}$$

where the upper limit, by analogy with the lower limit in (12.18), should be interpreted as meaning the value of z to the right of BB' above which there is negligible contribution to the integrand. As $d(z)$ increases, both β_+ and β_- rapidly approach the value of the fundamental-mode propagation constant β for either core in isolation. Clearly, on comparing (12.18) and (12.22), or by symmetry of the input and output cores, $\phi_o = \phi_i$.

12.7.7   Splitting ratio

Given the total accumulated relative phase change between the two supermodes over the complete length of the coupler, $\phi_{\text{tot}}(\lambda)$, where

$$\phi_{\text{tot}}(\lambda) = \phi_i + \phi_c + \phi_o = 2\phi_i + \phi_c \tag{12.23}$$

and ϕ_i, ϕ_c and ϕ_o are evaluated from (12.18),(12.20) and (12.22), respectively, the splitting ratio is readily determined. Recalling that unit power enters port 1 of Figure 12.4, it follows from (12.17), (12.19a) and (12.21a), that the amplitudes of the local supermodes sufficiently far to the right of BB' in Figure 12.4 are equal, i.e.

$$|c_+| = |c_-| \tag{12.24a}$$

and there is a relative phase between the modal constants, expressible as

$$c_- N_-^{1/2} = c_+ N_+^{1/2} e^{-i\phi_{\text{tot}}} \tag{12.24b}$$

Hence, on expressing ψ_+ and ψ_- in terms of ψ_1 and ψ_2 through (12.16), and suppressing the common phase and normalization N of the fundamental modes in cores 1 and 2, then the fields in the output arms 2 and 3 of Figure 12.4 are proportional to

$$(1 + e^{-i\phi_{\text{tot}}(\lambda)})\psi_1, \quad (1 - e^{-i\phi_{\text{tot}}(\lambda)})\psi_2 \tag{12.25}$$

respectively. Consequently, the corresponding output powers P_2 and P_3 are proportional to the squares of the moduli in (12.25), since ψ_1 and ψ_2 both have the same functional form. Imposing power conservation, so that the total output and input powers are equal, leads to

$$P_2 = \frac{1 + \cos \phi_{\text{tot}}(\lambda)}{2}, \quad P_3 = \frac{1 - \cos \phi_{\text{tot}}(\lambda)}{2} \tag{12.26}$$

so that the splitting ratio depends solely on the total accumulated relative phase change between the even and odd supermodes over the entire length of the coupler, including the curved input and output arms.

12.8 SYMMETRIC SINGLE-MODE COUPLERS

12.8.1 3 dB couplers

A single-mode planar coupler will act as a 3 dB splitter at a given input wavelength λ, with equal power outputs exiting from ports 2 and 3, provided that the total accumulated phase (12.23) satisfies

$$\phi_{\text{tot}}(\lambda) = \frac{2m+1}{2}\pi \tag{12.27}$$

for $m = 0, 1, 2, \ldots$. This is a transcendental equation which is solved numerically to determine the wavelength and coupler parameter values for which equal splitting can occur, and the shortest such coupler then corresponds to $m = 0$.

12.8.2 Wavelength multiplexing/demultiplexing couplers

For a coupler which will demultiplex and, by reciprocity, multiplex signals at 1.3 and 1.55 μm wavelengths, say, it is necessary that the superposition of the two signals entering port 1 in Figure 12.4 be split such that one wavelength exits from port 2 and the other from port 3. In other words, the coupler parameters must take values that simultaneously satisfy the two conditions

$$\phi_{\text{tot}}(1.3) = 2p\pi, \quad \phi_{\text{tot}}(1.55) = (2q+1)\pi \tag{12.28}$$

where p and q are positive integers. As with (12.27), these conditions are determined numerically. The values of the integers p and q need to be kept as small as possible to minimize the length of the coupler. This is also desirable to minimize any polarization effects due to non-square cross-section of the coupler, and also to maximize the yield of planar devices in any fabrication process.

12.8.3 Broadband 100% couplers

In Section 12.4, the expressions for the output powers P_2 and P_3 emerging from ports 2 and 3, respectively, given by (12.1), are quasi-periodic in wavelength λ. The dependence of the coupling coefficient on λ is not linear. Typical plots of P_2 and P_3 are shown in Figure 12.5.

In the neighbourhood of 100% coupling at wavelength $\lambda = \lambda_c$, say, the spectral dependence of the coupled power P_3 varies approximately quadratically

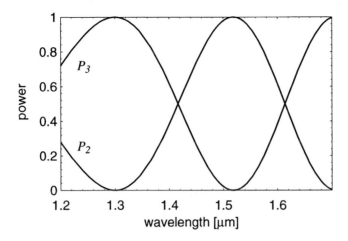

Figure 12.5 Quasi-periodic variation of the power outputs P_2 and P_3 of a symmetric coupler as a function of source wavelength.

according to

$$P_3 = 1 - \kappa(\lambda - \lambda_c)^2 \qquad (12.29)$$

where the constant κ is proportional to $(\partial C/\partial\lambda)^2$, evaluated at $\lambda = \lambda_c$. Now the wavelength dependence of the coupling coefficient has been shown to reach a maximum at a particular wavelength, λ_{max}, say (Ankiewicz *et al.*, 1986). If this maximum coincides with the 100% coupling wavelength, λ_c, then $\kappa = 0$ and the spectral dependence of P_3 now varies as (Love and Steblina, 1994)

$$P_3 = 1 - \sigma(\lambda - \lambda_c)^4 \qquad (12.30)$$

where σ is a constant proportional to $(\partial^2 C/\partial\lambda^2)^2$ evaluated at $\lambda = \lambda_c = \lambda_{max}$. Hence, the 100% coupling region is significantly broadened.

12.9   WAVELENGTH-FLATTENED AND BROADBAND COUPLERS

Symmetric, single-mode couplers, when used as 3 dB splitters, split signals equally at specific wavelengths. However, these devices have a maximum wavelength sensitivity precisely at the 3 dB splitting wavelength because of the periodic 100% power coupling between the two cores, corresponding to the crossing points of two curves in Figure 12.5. To reduce this sensitivity to wavelength while retaining the 3 dB split, slightly asymmetric fibre couplers have

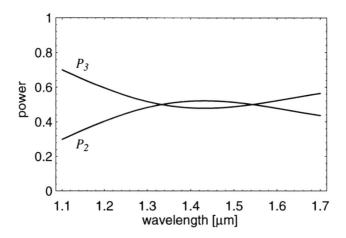

Figure 12.6 Power outputs P_2 and P_3 of the slightly asymmetric single-mode coupler.

been devised and developed (Mortimore, 1985). These couplers provide approximately 3 dB splitting over either the 1.3 or 1.55 μm windows, sometimes referred to as wavelength-flattened couplers, or over both windows, where they are sometimes referred to as wideband couplers. Asymmetric BCW couplers have also been analysed, designed and fabricated with spectral responses which are comparable to the corresponding fused-taper fibre couplers (Takagi *et al.*, 1992).

The basic principle behind both the wavelength-flattened and broadband couplers, is to detune the symmetric coupler slightly by introducing an asymmetry between the two cores, so that around 50% of the input power entering core 1 is coupled to core 2. This is equivalent to setting $F^2 \approx 0.5$ in the expressions for the power in each core of the asymmetric coupler in (12.11). Plots of the expressions for P_2 and P_3 as a function of wavelength are shown in Figure 12.6. Compared with the corresponding plots in Figure 12.5, the two curves now overlap slightly rather than cross at the 3 dB split, thereby significantly broadening the wavelength range over which approximately equal splitting occurs. For practical wavelength-flattened and broadband couplers, the fraction of power coupled is normally designed to be slightly greater than 50%. In this case, the curves for P_2 and P_3 cross over a narrow band of wavelengths around the operating wavelength, as illustrated in Figure 12.6; the net effect is to broaden the approximate 3 dB splitting region.

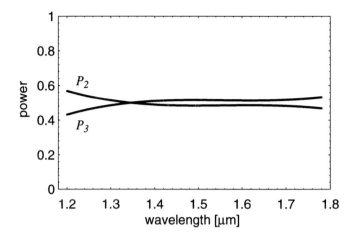

Figure 12.7 Plot of the power outputs for the highly-broadband asymmetric coupler. Coupler parameters are: core half-widths $\rho_1 = 3.6\ \mu$m and $\rho_2 = 2.3\ \mu$m, common height $2\rho_1 = 7.2\ \mu$m, separation between cores $d = 7.08\ \mu$m, cladding index 1.447, $\Delta = 0.003$ and central coupling region length 0.23 mm. The design includes the curved input/output arms of constant bend radius $R_c = 22$ mm. Estimated bend loss is approximately 0.1 dB for the entire coupler.

12.9.1   Highly-broadband 3dB coupling

In the neighbourhood of the wavelength λ_f in Figure 12.6, where the curves for P_2 and P_3 reach their extrema, P_3 differs from 0.5 with an approximately parabolic dependence on $\lambda - \lambda_f$. However, this range can be significantly broadened, and the coupling variation simultaneously reduced, by taking advantage of the wavelength maximum which occurs in the argument Cz_L/F of the expressions in (12.11).

Compared with the corresponding expressions for the symmetric coupler in (12.1), the effect of the additional factor, $1/F$, is to shift the peak value of the argument away from the wavelength λ_o, corresponding to the maximum in C. In the neighbourhood of the shifted peak, $\sin^2(Cz_L/F)$ has a parabolic variation with wavelength, while F^2 is monotonic. By a judicious choice of parameter values, it is possible to combine the parabolic dependence of $\sin^2(Cz_L/F)$ with the monotonic increase in F^2, to produce the characteristic cubic dependence of the powers P_2 and P_3 shown in Figure 12.7 for a highly-broadband BCW coupler. It has a variation of only 1.6% in the 50% coupling ratio over the range 1.3–1.67 μm (Love and Steblina, 1994).

REFERENCES

Ankiewicz, A., Snyder, A. W. and Zheng, X. H. (1986) Coupling between parallel optical fiber cores: critical examination. *IEEE Journal of Lightwave Technology*, **LT–4**, 1317–1323.

Hewlett, S. J., Love, J. D. and Steblina, V. V. (1995) Analysis and design of highly-broadband planar evanescent couplers. *Optical and Quantum Electronics*, **27**, in press.

Love, J. D. and Ankiewicz, A. (1985) Modal cutoffs in single- and few-mode fiber couplers. *IEEE Journal of Lightwave Technology*, **LT–3**, 100–110.

Love, J. D. and Steblina, V. V. (1994) Highly broadband buried channel couplers. *Electronics Letters*, **30**, 1853–1855.

Mortimore, D. B. (1985) Wavelength-flattened fused couplers. *Electronics Letters*, **21**, 742–743.

Snyder, A. W. and Love, J. D. (1983) *Optical waveguide theory.* London: Chapman & Hall.

Takagi, A., Jinguji, K. and Kawachi, M. (1992) Wavelength characteristics of (2×2) optical channel-type directional couplers with symmetric or nonsymmetric coupling structures. *IEEE Journal of Lightwave Technology*, **LT–10**, 735–746.

13

X-junctions

Propagation losses in single-mode, silica-based BCWs have now been reduced to such low values, typically of the order of a few dB per metre, that the excess loss across devices incorporated into BCW circuits should be minimal and not contribute significantly to overall loss. One device that has been developed is the right-angle X-junction, comprising two identical single-mode BCWs intersecting at 90° (Beaumont *et al.*, 1994). Such crossings help overcome the topographical constraint, which limits circuitry to a single plane for most current fabrication processes. These junctions have application to complex devices, such as networks with many input and output channels and intermediate devices which need to cross one another with minimal crosstalk. They also find use in long-length BCWs for sensing and nonlinear applications, where a very long length of BCW can be incorporated within a small area through spiralling. The X-junctions enable the inner end of the spiral to traverse itself to the outside without looping back on itself.

In this chapter, we determine the excess loss for the right-angle, single-mode BCW X-junction using both a simple approximation method (Love and Ladouceur, 1992; Ladouceur and Love, 1992) which generates an analytical expression for the excess loss, and the beam propagation method (BPM), to obtain accurate results and quantify the accuracy of this approximate solution. We also investigate the acute-angle junction, which may be more useful in situations where a 90° junction is inappropriate, and use the BPM to show that the excess loss remains very low until the intersection angle is small enough for cross talk between the two BCW cores to become the dominant loss mechanism.

13.1 MATHEMATICAL MODEL

Figure 13.1a shows a plane view of the right-angle, or 90° X-junction. Each of the four arms consists of square-core, single-mode BCWs of core width and height 2ρ, so that the dimensions of the intersection region are 2ρ by 2ρ as shown, with a common height throughout the device of 2ρ. Each BCW has a step profile with uniform core index n_{co} surrounded by an infinite cladding with uniform index n_{cl}, so that the intersection region also has index n_{co}. The choice of any other profile could lead to some indeterminacy of the profile in

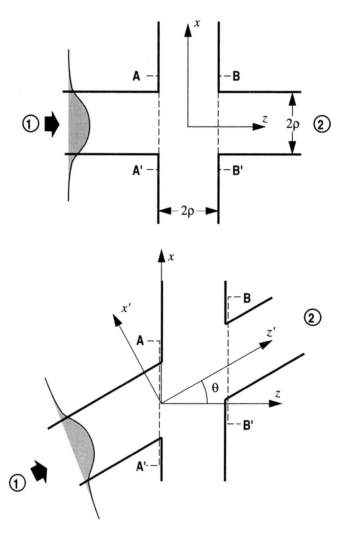

Figure 13.1 Parameters and geometry in the plane of (a) the right-angle *X*-junction and (b) the acute-angle *X*-junction.

the intersection and also increase excess loss. The arrangement for the acute-angle junction in Figure 13.1b is similar, where increasing the angle θ gives the departure from the 90° case. In either case, the fundamental mode of the BCW is assumed to enter the device through port 1, propagate across the intersection between AA′ and BB′ and leave through port 2.

13.2   PHYSICAL MODEL

When the fundamental-mode field entering the 90° junction through port 1 reaches the junction at AA′ in Figure 13.1a, it is no longer confined in the lateral, or x-direction, but remains confined in the vertical, or y-direction out of the plane of the figure. Accordingly, the modal field spreads out predominantly in the x-direction due to diffraction effects as it propagates across the width of the intersecting BCW. The diffracted beam then re-excites the fundamental mode over the cross-section BB′, and the remaining power in the beam is lost as radiation into the cladding.

The basic physical description of propagation across the acute-angle junction in Figure 13.1b is similar to that for the 90° junction, but is complicated by the skewed input port 1. Because the phase front of the mode is orthogonal to the direction of propagation, the tail of the fundamental-mode field in the cladding of the input arm will overlap the core of the intersecting BCW before the core part of the field reaches the intersection at AA′. Whilst the BPM simulation takes account of this effect, the simpler, analytical approximation discussed below does not.

At the intersection AA′ the field of the incident fundamental mode propagating in the cladding will be partially reflected because of the abrupt change in index from n_{cl} to n_{co}. Similarly, that part of the diffracted mode outside of the core at BB′ will also be partially reflected because of the change in index from n_{co} to n_{cl}. However, as we showed in Section 7.1.2, the relative loss due to this effect is of the order of $[(n_{co}-n_{cl})/2n_{co}]^2$, which, because of the slight difference in the core and cladding indices, is neglected in the following analysis.

13.3   APPROXIMATE ANALYSIS

The scalar fundamental-mode field of the square-core BCW has arbitrary polarization and the spatial dependence

$$e(x,y) = \Psi(x,y)e^{i\beta z} \qquad (13.1)$$

where $\Psi(x,y)$ denotes the transverse two-dimensional dependence, β is the propagation constant, and the z-direction corresponds to the axis of the BCW in port 1 in Figure 13.1a. Our simplified analysis approximates the field of the mode propagating using the Gaussian approximation of Section 4.4.

Across the intersection, the spread of the modal field is described by Fresnel diffraction, since the spot size s of the field cannot be neglected compared with $\sqrt{(2\rho/k)}$, which is the delineation criterion between Fresnel and Fraunhofer diffraction, where 2ρ is the width of the crossing region and k is the free-space wavenumber. Further, we assume that the field is confined in the y-direction perpendicular to the plane of the junction and that it diffracts only in the transverse direction. The diffracted beam then re-excites the outgoing fundamental mode on the opposite side of the junction. The amplitude of this mode is also calculated within the Gaussian approximation.

Under these assumptions, the diffracted field is determined by the Fresnel diffraction integral (Hauss, 1984), so that in the region between AA$'$ and BB$'$, the complete spatial dependence of the field is given by

$$\Psi_d(x,y,z) = \frac{ikn_{co}}{2\pi z} \int_{-\infty}^{+\infty} dx_0 \int_{-\infty}^{+\infty} \Psi(x_0,y_0) e^{-i\left(\frac{kn_{co}}{2z}\right)\left[(x-x_0)^2 + (y-y_0)^2\right]} dy_0$$

(13.2)

where x_0 and y_0 denote the transverse coordinates at the beginning of the junction at AA$'$ in Figure 13.1a, $k = 2\pi/\lambda$ and λ is the source wavelength. Within the Gaussian approximation, the modal field is separable in x and y, so we can reduce the X-junction to a one-dimensional problem. If we define the modal field as the product of x and y spatial dependencies, $\Psi(x,y) = \psi(x)\phi(y)$, then (13.2) reduces to

$$\Psi_d(x,y,z) = \phi(y)\sqrt{\frac{ikn_{co}}{2\pi z}} \int_{-\infty}^{+\infty} \psi(x_0) e^{-i\left(\frac{kn_{co}}{2z}\right)(x-x_0)^2} dx_0 \qquad (13.3)$$

If we further assume that the y-dependence of the problem remains invariant across the junction, then (13.3) reduces to

$$\psi_d(x,z) = \sqrt{\frac{ikn_{co}}{2\pi z}} \int_{-\infty}^{+\infty} \psi(x_0) e^{-i\left(\frac{kn_{co}}{2z}\right)(x-x_0)^2} dx_0 \qquad (13.4)$$

for the one-dimensional problem.

13.4   RIGHT-ANGLE **X**-JUNCTION

The fundamental-mode field Ψ of the square-core waveguide is well approximated by the circularly-symmetric Gaussian field presented in Section 4.4:

$$\Psi(x,y) = \exp\left(-\frac{x^2 + y^2}{2s^2}\right) \qquad (13.5)$$

where the spot size s is determined using the standard variational procedure of Section 4.4. We use the separability of the field to define the transverse

x-dependence of the modal field

$$\psi(x) = \exp\left(-\frac{x^2}{2s^2}\right) \tag{13.6}$$

with corresponding normalization N

$$N = \int_{-\infty}^{+\infty} |\psi(x)|^2 \, dx = \sqrt{\pi}s \tag{13.7}$$

The diffracted field is calculated by substituting (13.6) into the Fresnel diffraction integral (13.4). Straightforward integration leads to

$$\psi_d(x, z) = \frac{s}{(s^2 - i\sigma^2)^{1/2}} \exp\left[-\frac{x^2}{2(s^2 - i\sigma^2)}\right] \tag{13.8}$$

where $\sigma = \sqrt{z/k_{co}}$. The complex argument in the exponential factor accounts for both the transverse spread of the modal field and associated curvature of the initially plane phase front at AA$'$ as it propagates across the intersection. The fraction of power recaptured by the fundamental mode in port 2of Figure 13.1a at BB$'$ is given by the square of the overlap integral, I, of the diffracted field and the modal field, where(13.6)

$$I(z) = \frac{1}{N} \int_{-\infty}^{+\infty} \psi_d(x, z)\psi(x)^* \, dx \tag{13.9}$$

and $*$ denotes complex conjugate. This integration is performedat $z = 2\rho$, and the transmission coefficient for the fraction of incident power in port 1 propagating along port 2 is readily found to be

$$T = |I|^2 = \frac{kn_{co}s^2}{(kn_{co}^2 s^4 + \rho^2)^{1/2}} \tag{13.10}$$

Using $V = kn_{co}\rho\sqrt{2\Delta}$ to replace kn_{co} by the BCW V-value, and noting that $kn_{co}s^2 \gg \rho$ for the BCWs, since s and ρ are approximately equal and $kn_{co} \gg 1$, the fractional power loss $L = 1 - T$ is well approximated by

$$L = 1 - T = \frac{\Delta}{V^2}\left(\frac{\rho}{s}\right)^4 \tag{13.11}$$

where Δ is the relative index difference. Since the spot size s and the core half-width ρ are similar for single-mode BCWs with $V = 2$, the loss depends mainly on the relative-index difference. In other words, for X-junctions with low index difference, the corresponding excess loss will be small.

13.4.1   Heuristic derivation

An alternative derivation to the formal development given the previous section provides a heuristic approach to the loss formula (13.11) in terms of the basic physical processes involved. We assume that the fundamental mode entering port 1 in Figure 13.1a diffracts at an angle θ_d relative to the z-direction of propagation at the beginning of the intersection AA'. As the fundamental-mode field distribution in the x-direction is approximately Gaussian, one-dimensional diffraction theory for a Gaussian beam in the far field, or Fraunhofer diffraction limit gives (Snyder and Love, 1983, Chapter 10)

$$\theta_d = \frac{\lambda}{2\pi n_{co}s} \tag{13.12}$$

in terms of the source wavelength λ. By geometry, the spot size s increases from s to $s + 2\rho\theta_d$ in propagating across the junction. To determine the fraction of power which is radiated at BB', we use the overlap integral for two Gaussian fields with spot sizes s_1 and s_2 given by (Snyder and Love, 1983, Chapter 20)

$$\left(\frac{2s_1 s_2}{s_1^2 + s_2^2}\right)^{1/2} \tag{13.13}$$

By setting $s_1 = s$ and $s_2 = s + 2\rho\theta_d$, it is straightforward to show that the fractional power loss is given by

$$L = 4\frac{\Delta}{V^2}\left(\frac{\rho}{s}\right)^4 \tag{13.14}$$

This expression is identical to (13.11) in its parametric dependence, but differs numerically from it by a factor of 4. The difference is attributable to two factors. Firstly, we used the Fraunhofer limit for the diffraction angle, and this is significantly larger than the value obtained in the Fresnel régime used in the previous section, and secondly, we have ignored the effect of curvature of the phase fronts in the analysis.

13.5   ACUTE-ANGLE *X*-JUNCTION

The analysis presented in the previous section is readily adapted to the case where the angle of the X-junction deviates from 90°. The basicassumptions described in Section 13.2 are retained, and we also assume that the modal field at the beginning of the X-junction, i.e. in the plane of the cross-section AA' in Figure 13.1b, is unperturbed by the transverse waveguide. In other words, in the coordinate system shown, the input modal field entering the X-junction from port 1 is approximated by a Gaussian field of the form

$$\Psi(x', y', z') = \exp\left(-\frac{x'^2 + y'^2}{2s^2}\right)\exp(i\beta z') \tag{13.15}$$

Clearly, this assumption is only valid for a small values of θ in Figure 13.1b. As explained above, we can suppress the y'-dependence as the field, in which case the Gaussian field on $z = 0$ relative to the (x, z) coordinate system is now given by

$$\psi(x,z)|_{z=0} = \exp\left(-\frac{x^2 \cos^2 \theta}{2s^2}\right) \exp(i\beta x \sin \theta) \tag{13.16}$$

with the corresponding normalization

$$N = \sqrt{\pi}s \sec \theta \tag{13.17}$$

The diffracted field that propagates across the X-junction is obtained from the diffraction integral (13.4). This integral can be expressed in closed form using

$$\int_{-\infty}^{+\infty} dx \exp(-ax^2 + bx) = \sqrt{\frac{\pi}{a}} \exp\left(\frac{b^2}{4a}\right), \quad \mathrm{Re}(a) > 0 \tag{13.18}$$

This result, together with (13.4) and (13.16), leads to the diffracted field

$$\psi_d(x,z) = \frac{s}{\nu} \exp\left[-\frac{ix^2}{\sigma^2} - \frac{i\sigma^2 s^2}{2\nu^2}\left(\frac{ix}{\sigma^2} + i\beta \sin \theta\right)^2\right] \tag{13.19}$$

where $\nu = s^2 - i\sigma^2 \cos^2 \theta$. The fraction of power transmitted through the junction into the fundamental mode in port 2 is obtained by taking the overlap integral of the diffracted field $\psi_d(x,z)$ with the local mode of the waveguide in port 2 of the X-junction, i.e. with

$$\psi(x,z)|_{z=2\rho} = \exp\left[-\frac{(x-x_d)^2 \cos^2 \theta}{2s^2}\right] \exp(i\beta x \sin \theta) \tag{13.20}$$

where the $x_d = 2\rho \tan \theta$. We used the symbolic algebra package Mathematica (Wolfram, 1988) to carry out this last integration, as the amount of algebra becomes excessive and is thus prone to errors. Two approximations were also introduced:

$$\beta \sigma^2 = \frac{\beta}{k_{co}} 2\rho \approx 2\rho$$

$$\sigma/s \ll 1$$

where the former follows from weak-guidance, since $\beta \approx k n_{co}$, and the latter is essentially equivalent to $s \gg \lambda$. Within these approximations we finally obtain

$$T = \left[1 - \frac{\Delta}{V^2}\left(\frac{\rho}{s}\right)^4 \cos^4 \theta\right] \exp\left[-32\left(\frac{\rho}{s}\right)^2 \cos^2 \frac{\theta}{2} \sin^6 \frac{\theta}{2}\right] \tag{13.21}$$

The small deviation angle limit, $\theta \to 0$, simplifies this expression to

$$T = \left[1 - \frac{\Delta}{V^2}\left(\frac{\rho}{s}\right)^4\right] \exp\left[-\frac{1}{2}\left(\frac{\rho}{s}\right)^2 \theta^6\right] \tag{13.22}$$

This result shows that for a slight departure from the 90° junction, the additional excess loss incurred is very small.

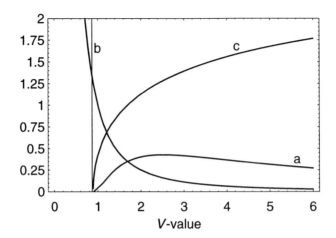

Figure 13.2 Plots of (a) $L/\Delta = (1/V)^2(1/S)^4$, (b) $1/V^2$ and (c) $1/S$ as functions of the BCW V-value, for constant profile height, Δ. Curve (a) reaches a maximum at $V = 2.52$.

13.6   CHARACTERISTICS OF THE APPROXIMATE MODEL

When using the approximate X-junction diffraction model, there are various attributes of the resulting algebraic expressions (13.11) and (13.22) which should be kept in mind (Hewlett *et al.*, 1994). Firstly, (13.11) dictates that the excess loss is a single-valued function of the waveguide's degree of guidance, V, when the profile height, Δ, remains constant, i.e. $L/\Delta = (1/V^2)(1/S^4)$, which is a function of the V-value only, because the normalized spot size S depends only on V. This result is quantified in Figure 13.2, and shows that the factor $1/V^2$ (curve b) is monotone decreasing and the factor $1/S$ (curve c) is monotone increasing with increasing V-value. Consequently, L/Δ (curve a) reaches a maximum value at $V = 2.52$.

Thus, increasing the degree of guidance can lead to either an increase or a decrease in excess loss, depending on which side of the extremum the X-junction is being operated. The beam propagation method (BPM) simulations presented in the following section confirm this behaviour and indicate that the analytical result of (13.11) accounts for the dominant physical processes involved in the diffraction process. This result, with slight modification, has been used to explain the loss incurred in high-Δ waveguide X-junctions (MacKenzie *et al.*, 1992).

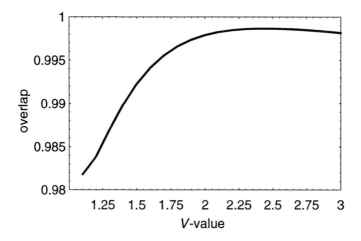

Figure 13.3 Plot of the overlap integral between the Gaussian approximation of the BCW fundamental mode and the exact BPM-calculated modal field as a function V. The value of V was varied by changing the wavelength, other parameters maintaining the fixed values $\rho = 4\ \mu\text{m}$, $n_{\text{co}} = 1.45$ and $n_{\text{cl}} = 1.447$.

13.6.1   Accuracy of the Gaussian approximation

As shown in Figure 6.3, the underlying Gaussian approximation becomes increasingly inaccurate for V-values approaching $\sqrt{\pi}/2 \approx 0.886$. As pointed out in Section 4.4, the accuracy of the Gaussian approximation and hence that of the approximate results for the X-junction, is best for values of V around $V = 2$, which, fortuitously, lies in the range of practical interest, below the second-mode cutoff at $V = 2.14$. This accuracy is quantified in Figure 13.3, which plots the overlap between the Gaussian approximation of (13.5) using the numerically-evaluated spot-size from (4.30), and the exact fundamental-mode field calculated using the BPM. The figure caption details the parameter values used for these simulations. Unit overlap corresponds to perfect agreement between the approximate and exact modal fields. If a 1% discrepancy is taken as the maximum tolerable error, the Gaussian approximation holds for $V \geq 1.41$ in the range shown.

13.6.2   Series of X-junctions

The model assumes that the X-junction is considered in isolation, i.e. the four arms should be sufficiently long to allow the non-guided part of the fundamental-mode field to radiate sufficiently far away from the cores. If a closely-separated series of X-junctions is built into a BCW, i.e. when the distance separating the X-junction is comparable with the width of the X-junctions themselves, then recoupling of the radiation field to the guided mode can occur, leading to possible interference with the fundamental-mode signal. Resonance effects can also be set up by the grating-like periodic structure formed by cascaded X-junctions. A more appropriate model should be adopted in this case (Capobianco *et al.*, 1993).

13.6.3   Diffraction out of the X-junction plane

The approximate model assumes there is no diffraction in the y-direction, i.e. out of the plane of the X-junction. This approximation limits the overlap between the diffracted and undiffracted fields, thereby underestimating the excess loss. Hence, (13.11) and (13.22) should be regarded as approximate lower bounds on the loss. Now consider the right-angle X-junction when the diffraction in the y-direction is also taken into account. If there were no guidance in the y-direction, then following the derivation of Section 13.4 the overall loss would be twice that given by (13.11). Clearly the actual loss is significantly less than this value because of the guidance across the intersection, but it provides an upper bound on the total X-junction loss. Thus, the actual loss L, for the right-angle X-junction is loosely bounded by

$$\frac{\Delta}{V^2} \left(\frac{\rho}{s} \right)^4 < L < 2\frac{\Delta}{V^2} \left(\frac{\rho}{s} \right)^4 \qquad (13.23)$$

13.7   NUMERICAL RESULTS

A full two-dimensional scalar beam propagation method (BPM) simulation (Feit and Fleck, 1978) was used to determine excess loss numerically by propagating the exact fundamental mode of the BCW across both the 90° and acute-angle X-junctions. The loss is determined by overlapping the propagated beam sufficiently far a long port 2 with the input beam in port 1 to ensure that the power radiated from the intersection is far away from the core. For the right-angle junction, a 512×512 discretization grid spanning a physical domain of $102.4~\mu\text{m} \times 102.4~\mu\text{m}$, together with a propagation step of $0.2~\mu\text{m}$, was found to be sufficiently accurate. For the acute angle, a 256×128 discretization grid spanning a physical domain of $128~\mu\text{m} \times 64~\mu\text{m}$ was used; the other parameters being unchanged.

For the approximate analytical solution of Section 13.4, the spot size for the Gaussian approximation to the square-core BCW fundamental mode is given

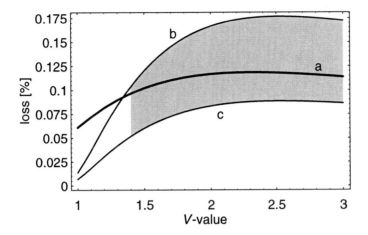

Figure 13.4 Plots of (a) the BPM-calculated percentage power loss, (b) upper bound and (c) lower bound loss values obtained from (13.23) as a function of BCW V-value. The shaded region shows the range of validity of the Gaussian approximation, based on the maximum 1% overlap integral error in Figure 13.3.

in terms of the normalized spot size $S = s/\rho$, determined from the numerical solution of the eigenvalue equation (4.30), and presented in Table 9.1.

13.7.1   Right-angle X-junction

The exact excess loss calculated from the BPM simulation is plotted against the BCW V-value, as a percentage of power in the input fundamental mode, as curve a in Figure 13.4. The simulation parameters are the same as those used for Figure 13.3. Curves b and c correspond, respectively, to the upper and lower bounds on L given in (13.23). Curve a exhibits a maximum near the value $V = 2.52$ predicted from the approximate solution. This maximum may account for experimental observations in the 90° junction of decreasing loss with increasing V (MacKenzie *et al.*, 1992), and in acute-angle junctions (Aretz and Bülow, 1989; Murphy *et al.*, 1994), where both the excess loss and crosstalk values were significantly reduced by either adiabatically up- or down-tapering the waveguides before they intersect.

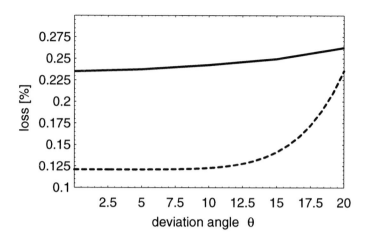

Figure 13.5 Percentage excess loss L as a function of the deviation θ from the 90° junction for a BCW with $V = 2$. The solid curve is the BPM result and the dashed curve is the analytical approximation $L = 1 - T$ calculated from (13.22).

13.7.2   Acute-angle **X**-junction

Figure 13.5 plots the percentage loss, L, as a function of the angle θ, with increasing values of θ corresponding to the departure from the 90° junction. The solid curve denotes the exact BPM result, and the dashed curve is calculated from (13.22), assuming a $V = 2$ for both curves. There is a relative error of less than 50% in the approximate result, and both curves exhibit similar behaviour up to about 10° deviation from the right-angle case.

The solid curve in Figure 13.6 shows the BPM calculation of excess loss for $V = 2$ over a range of deviation angles θ spanning 0° to 85°. Clearly excess loss remains very small even for angles up to 70°, corresponding to a 20° intersection angle between the BCWs. This should enable designs of optical circuitry to use the available area more efficiently as optical lines can cross over a large range of angles with minimal loss. As the deviation angle approaches 90°, the junction behaves more like a coupler and the analysis must account for crosstalk, discussed in the following section. The dashed curve is the prediction of (13.21) showing the rapid departure from the exact solution for larger values of θ. This is due to the more complicated interaction between the fundamental mode field in the input and the core of the intersecting skew BCW.

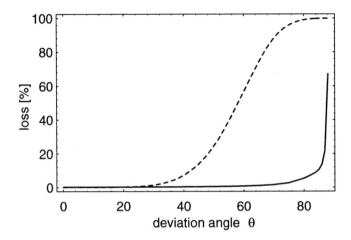

Figure 13.6 Extension of the results shown in Figure 13.5 to a broader range of crossing angles θ.

13.7.3   Crosstalk

The excess loss for the acute-angle junction plotted in Figure 13.6 becomes significant only for crossing angles between the two BCWs which are below about 10°. However, as $\theta \to 90°$, the dominant loss mechanism is crosstalk, or coupling between the fundamental mode in port 1 in 13.1 and the fundamental mode in the the cross-BCW.

We can obtain a simple criterion for the onset of crosstalk by assuming that coupling occurs only if the modal phase velocity of one BCW relative to the other lies in the range for bound-mode propagation. Thus, if β is the propagation constant of the fundamental mode propagating through port 1, then the corresponding phase velocity parallel to the axes is ω/β, where ω is the source frequency. Relative to the axes of the skew BCW, the phase velocity is $\omega \cos\phi$, where $\phi = \pi/2 - \theta$ is the angle between the axes of the two BCWs. Thus the propagation constant along the skew guide is $\beta/\cos\phi$, which, if the fundamental mode is to propagate cannot exceed the upper limit $2\pi n_{co}/\lambda$. On rearranging this expression in terms of the BCW and modal parameters, we find, that for small values of ϕ, crosstalk will occur if

$$\phi < (2\Delta)^{1/2}\frac{U}{V} \tag{13.24}$$

where $U = \rho(k_{co}^2 - \beta^2)^{1/2}$. For example, this expression gives $\phi < 3.2°$ for a

BCW with $\Delta = 0.3\%$ and $V = 2$ (corresponding to $U = 1.446$). This is consistent with the position of the steepest part of the solid curve in Figure 13.6.

REFERENCES

Aretz, K. and Bülow, H. (1989) Reduction of crosstalk and losses of intersecting waveguides. *Electronics Letters*, **11**, 730–731.

Beaumont, C. J., Cassidy, S. A., Welbourn, D., Nield, M. and Thurlow, A. (1994) Integrated silica optical delay line. Pages 241–244 of: *Proceedings of the 17th European Conference on Optical Communication.*

Capobianco, A. D., Costantini, B. and Someda, C. G. (1993) BPM modelling of planar right-angle X junctions. *Electronics Letters*, **29**, 753–755.

Feit, M. D. and Fleck, J. A. Jr. (1978) Light propagation in graded-index optical fibers. *Applied Optics*, **17**, 3990–3998.

Haus, H. A. (1984) *Waves and fields in optoelectronics.* Englewood Cliffs, New Jersey: Prentice-Hall.

Hewlett, S. J., Ladouceur, F. and Love, J. D. (1994) On the right-angle *X*-junction diffraction model. *IEE Optoelectronics Part-J*, to be published.

Ladouceur, F. and Love, J. D. (1992) *X*-junctions in buried channel waveguides. *IEE Optoelectronics Part-J*, **24**, 1373–1379.

Love, J. D. and Ladouceur, F. (1992) Excess loss in singlemode right-angle *X* junctions. *Electronics Letters*, **28**, 221–222.

MacKenzie, F., Beaumont, C. J., Nield, M. and Cassidy, S. A. (1992) Measurement of excess loss in planar silica X junctions. *Electronics Letters*, **28**, 1919–1920.

Murphy, T. O., Irvin, R. W. and Murphy, E. J. (1994). Uniform low-loss waveguide interconnects. Pages 184–186 of: *Proccedings Integrated Photonics Research.*

Snyder, A. W. and Love, J. D. (1983) *Optical waveguide theory.* London: Chapman & Hall.

Wolfram, S. (1988) *Mathematica: a system for doing mathematics by computer.* Redwood city: Addison-Wesley.

14

Y-junctions

An important passive device in planar optical technology is the single-mode, symmetric planar Y-junction, illustrated schematically in Figure 14.1. It consists of a stem and two arms, each comprising a single-mode, square-core BCW. Since the vertical thickness of the device is constant, as dictated by the etching-deposition fabrication process, its cross-section will be rectangular in the region where the stem and arms meet. Intuitively, the angle between the two arms should be kept small in order to minimize the excess loss due to the effective uptapering of the arms and to ensure that the device is approximately adiabatic.

The functionality of the Y-junction relies solely on its symmetry. If the fundamental mode of the stem is excited, it will split equally between the two arms and, even allowing for excess loss, the optical power exiting the two arms will be equal. In other words, the device is nominally a 3 dB splitter. Furthermore, since the splitting action depends only on symmetry, it is independent of wavelength. Thus the Y-junction is a wavelength-independent 3 dB splitter.

Conversely, when the Y-junction is operated in the reverse direction, the

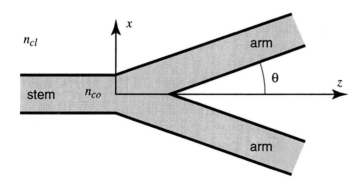

Figure 14.1 Schematic representation of a simple Y-junction consisting of a straight stem, a linear taper section and straight arms.

fundamental mode of either arm is excited and propagates through the junction to become the fundamental mode of the stem, but with a 50% loss of power, even if the device is adiabatic when operated in the forward direction. The 50%, or 3 dB loss is also independent of wavelength, and is a consequence of the reciprocity property of the *Y*-junction. An alternative explanation of this property, in terms of mode propagation, is given in the next section.

The emphasis in this chapter is on the design of an optimally low-loss *Y*-junction, using simple physical principles, and constrained by parameter values dictated by telecommunication applications, and by the limitations imposed by the lithographical and fabrication processes. In the following sections, we present an analysis of the tapered and splitting sections in order to develop a complete low-loss *Y*-junction design (Ladouceur, 1991). Section 14.4 covers the constraints that lead to the design adopted for numerical simulation.

14.1   MODAL MODEL

14.1.1   Forward propagation

Propagation in either direction through the single-mode *Y*-junction can be explained in terms of its modes. As the stem is single-moded, it supports only the fundamental mode. If the stem is excited, this mode propagates in the positive z-direction in Figure 14.2a. When it reaches the junction with the two arms, provided the change in the core cross-section is slow enough, as discussed in the following section, this mode follows the uptaper approximately adiabatically, conserving virtually all its power, retaining its symmetry and adapting its modal field distribution to match the change in cross-section of the core. In the region beyond the junction, the modal field has the symmetric two-humped distribution shown in Figure 14.2a, i.e. it is the even, or fundamental supermode discussed in Section 14.5. Furthermore, the supermode is essentially the local supermode because of the approximate conservation of power. Because of the symmetry, there must be equal power in each core, and as the above description does not involve the source wavelength, the equal-splitting action is independent of wavelength.

Power is conserved only if the local supermode is adiabatic, i.e. it does not couple to the continuum of higher-order radiation modes. In practice there will always be some excess loss because of the uptaper, but as we show in later sections, this can be minimized by judicious choice of design parameters.

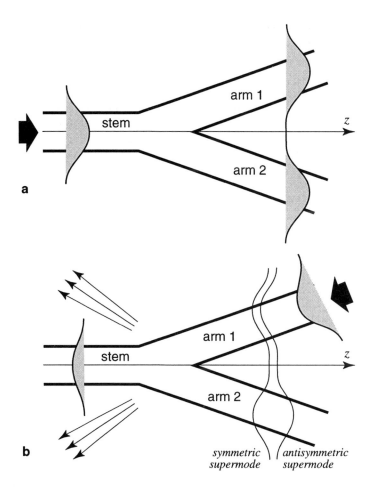

Figure 14.2 Schematic representation of the evolution of modes of the Y-junction when excited (a) along the stem, and (b) through one of the arms.

14.1.2   Backward propagation

If one of the arms of the Y-junction is excited, as in Figure 14.2b, the incoming fundamental mode in the upper arm is equal to the superposition of the even and odd local supermodes of the composite two-core waveguide defined by the two arms. As the even and odd supermodes correspond approximately to the sum and difference, respectively, of the fundamental-mode fields in each arm, each local supermode carries essentially half of the power of the fundamental mode in the upper arm.

The even local supermode propagates approximately adiabatically through the junction of the two arms, and becomes the fundamental mode in the stem, whereas the odd supermode cannot be supported beyond the junction because the stem is single moded and therefore must radiate all its power. Hence only 50% of the power in the arm is transmitted to the stem. This figure will be further reduced by the slight non-adiabaticity of the even local supermode.

14.2   TAPERED SECTION

Changing the cross-section of a single-mode waveguide by tapering necessarily induces radiation loss from the fundamental mode, because of the loss of translational invariance. In principle, loss can be kept arbitrarily small by making the taper angle everywhere sufficiently small, but such devices are difficult to fabricate and are impractically long. A more thorough examination of the variation of radiation loss with the rate of tapering reveals that there is an optimal taper shape which minimizes taper length for a given overall loss (Love and Henry, 1986). A simple physical interpretation of this result can be given in terms of coupling and taper length scales (Stewart and Love, 1985; Love, 1989).

14.2.1   Loss criterion

Consider the tapered single-mode fibre core shown in Figure 14.3 where z is the distance measured along the taper axis, $\rho(z)$ the core radius at position z, and $\Omega(z)$ is the local taper angle between the waveguide axis and the tangent to the core–cladding interface in the plane of the Y-junction. We define a taper length scale z_t by the length of the cone with base coincident with the core cross-section and semi-vertical angle $\Omega(z)$. Since $\Omega(z)$ is small, we have

$$z_t \approx \rho(z)/\Omega(z) \qquad (14.1)$$

Assuming that the cladding can be taken to be infinitely thick, the second length scale is the coupling length between the fundamental mode and the radiation field, which coincides with the beat length z_b. When theradiation is slight, it will be predominantly in the forward direction, parallel to the fibre axis, in which case

$$z_b = \frac{2\pi}{\beta(z) - kn_{cl}} \qquad (14.2)$$

where n_{cl} is the cladding index, k is the free-space wavenumber, and $\beta(z)$ is the fundamental-mode propagation constant at position z, i.e. the propagation constant for a uniform fibre with the profile and cross-sectional geometry at position z in Figure 14.3.

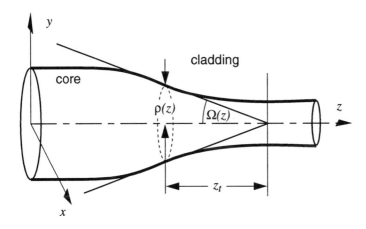

Figure 14.3 Schematic representation of a tapered single-mode fibre.

It is intuitive that radiation loss will be small if the taper length scale is large compared with the coupling length, and, conversely, radiation will be large if the taper length scale is quite short. Thus a simple delineation between the two extremes is obtained by equating the two length scales (Love, 1989), i.e.

$$z_b = z_t \tag{14.3}$$

On substituting from (14.2) and (14.3) and rearranging, we obtain

$$\Omega_d(z) = \frac{\rho(z)(\beta - kn_{cl})}{2\pi} \tag{14.4}$$

In other words, the delineating value of the local taper angle, $\Omega_d(z)$, is given implicitly in terms of the local core radius and profile through the local propagation constant.

14.2.2   Tapered BCW

Tapering of a BCW is achieved in the fabrication process by introducing a taper into the mask design. The application of the various deposition/etching stages to the mask taper results in a BCW whose core height in the vertical plane remains constant, while the horizontal width varies. Figure 14.4 shows such a tapered mask design. The cross-section BB′ corresponds to the ideal square-core cross-section with side $2\rho_x = 2\rho_y$, CC′ to a downtapered cross-section with width $2\rho_x < 2\rho_y$, and AA′ to an uptapered cross-section with width $2\rho_x > 2\rho_y$.

If we assume that the square cross-section BB′ corresponds to $\rho_x = \rho_y$ and $V = 2$, then, by introducing a generalized definition for the waveguide

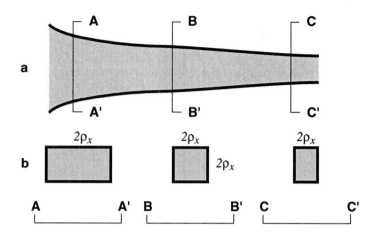

Figure 14.4 (a) Mask design for fabricating a tapered BCW, and (b) the core cross-section at the corresponding positions along (a).

parameter, viz.

$$V = k(\rho_x \rho_y)^{1/2}(n_{co}^2 - n_{cl}^2)^{1/2} \qquad (14.5a)$$
$$= k(\rho_x \rho_y)^{1/2} n_{co}(2\Delta)^{1/2} \qquad (14.5b)$$

the waveguide at cross-section AA′ has $V > 2$, and at CC′ it has $V < 2$. It is convenient to introduce a comparable normalized modal parameter W, defined by

$$W = (\rho_x \rho_y)^{1/2}(\beta^2 - k^2 n_{cl}^2)^{1/2} \qquad (14.6)$$

to determine the local propagation constant from the eigenvalue equation. If (14.5a) and (14.6) are substituted into (14.4), then the maximum permissible local taper angle for approximately adiabatic propagation is expressible as

$$\Omega_d = \frac{\sqrt{2\Delta}}{4\pi}\frac{W^2}{V} \qquad (14.7)$$

We use the FDM of Chapter 5 to evaluate W, as V varies with the change in ρ_x. Assuming $\Delta = 0.003$, (14.7) leads to the delineation curve plotted in Figure 14.5 for the value of normalized taper angle $4\Omega\pi/\sqrt{2\Delta}$ as a function of V. Thus $V = 2$ corresponds to the square cross-section in Figure 14.4, $V > 2$ to the uptapered core cross-section and $V < 2$ to the downtapered cross-section.

To design an approximately adiabatic taper, the value of the local taper angle must lie below the delineating curve in Figure 14.5. Values lying above this curve correspond to relatively large radiation loss. The dashed curve represents an approximately adiabatic taper, corresponding to the taper shape in Figure 14.4.

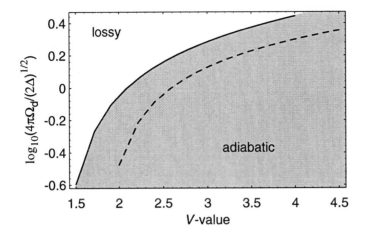

Figure 14.5 Normalized delineation angle $\log(4\pi\Omega_d/\sqrt{2\Delta})$ (solid curve) as a function of V of (14.5a). The white and shaded areas denote, respectively, the lossy and approximately adiabatic regions. The dashed curve corresponds to a low-loss tapered BCW design for operation in the adiabatic region.

This result will be used later in this chapter in the design of an optimally low-loss Y-junction for the uptaper from the single-moded stem to the two-moded region before the split in the arms at the 'vee' in Figure 14.1.

14.3   SPLITTING SECTION

14.3.1   Split angle

The divergence of the two arms of the Y-junction in Figure 14.1 is equivalent to an uptaper, which introduces excess loss through coupling of the local supermodes to the radiation field. Intuitively, the excess loss will increase with increasing the angle θ in the 'vee' of Figure 14.1. Thus a splitter with arbitrarily low loss could be designed by making this angle sufficiently small, but, as with the taper discussed above, this would lead to an unacceptably long device difficult to fabricate. Fortunately, it is possible to balance the twin requirements of low loss and compactness using simple physical considerations, similar to those applied to the taper in Section 14.2, and to obtain a simple upper bound on this angle (Love, 1989).

As we saw in Section 14.1, propagation along the arms can be expressed in terms of a superposition of the even and odd local supermodes. Thus, by following the development in Section 14.2.2, the maximum allowable splitting angle θ for approximately adiabatic propagation in Figure 14.3 would be given by (14.7) in terms of the cladding mode parameters W_+ and W_- for the even and odd supermodes, respectively, for the same value of V. As $W_- < W_+$, the odd supermode imposes the more severe constraint. However, the cross-sectional area of the BCW at the beginning of the arms in Figure 14.1 is double that of the stem and the waveguide is, therefore, two-moded. Consequently, the difference between the value of W_- at the 'vee' and W for the stem is not large, so the former can be approximated by the latter. Accordingly, it follows from (14.7) that

$$\theta < \frac{(2\Delta)^{1/2}}{4\pi} \frac{W^2}{V} \tag{14.8}$$

where θ is the half-branching angle. For representative values $V = 2.0$, $W = 1.45$ and $\Delta = 0.003$ for this single-mode BCW stem, the angle should be below one degree.

14.3.2   Bent arms

If the arms of the Y-junction consist of straight BCWs, then the divergence between the two arms increases linearly with the length of the device. Assuming that $\theta = 0.25°$, in order to satisfy the divergence constraint on the arms of (14.8), then, in order to splice the arms to the cores of standard 125 μm fibres, the Y-junction must be long enough for the two arms to be at least 125 μm apart. Simple geometry leads to a device at least 14 mm long, so that a 1 × 8 splitter comprising concatenated Y-junctions would be at least 42 mm long.

   A much more compact Y-junction, and, consequently, a more compact concatenation of Y-junctions can be designed by introducing curvature into the arms, so that initially they diverge away from one another, as illustrated in Figure 14.6. The curvature can subsequently be reversed to bring the arms parallel to one another at a separation appropriate for splicing to adjacent fibres or to a concatenation of two Y-junctions.

   However, in bending the arms of the Y-junction, the bend radius should be large enough to avoid introducing significant bend loss. Bend loss for BCWs has already been discussed and quantified in Chapter 10. For typical single-mode BCW parameters, a minimum bend radius of 10 mm is required, so that, initially, the arms of the Y-junction would not diverge much more rapidly than in the linear case. Chapter 11 also introduced various strategies for reducing bend loss.

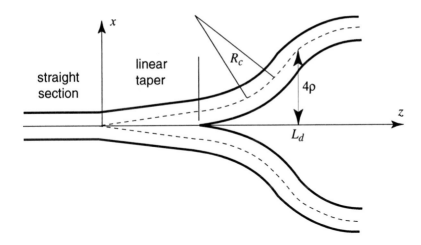

Figure 14.6 Schematic representation of a Y-junction consisting of a straight stem, a linear uptaper section and arms with constant radius of curvature R_c.

14.4 DESIGN CONSTRAINTS

The ideal compact Y-junction design incorporates the tapering and bending strategies described in the previous sections, but cannot be readily achieved in practice with great accuracy, due to various uncertainties associated with the fabrication process. One inaccuracy which can arise is in the mask design process. Current lithographical techniques for mask design and transfer have a resolution that is typically of the order of 0.05 μm, which means that the salient features of the Y-junction can only be specified with this tolerance.

The nature of the PECD deposition process is such that it is difficult to fill narrow etched regions with material, if the transverse dimension is of the order of a micron. This limitation inhibits the fabrication of a Y-junction with a well-defined 'vee' between the arms as illustrated in Figure 14.6. Accordingly a practical prototype design should include a U-shaped notch between the two arms with a width of the order of 1–2 μm. The introduction of this notch also helps reduce excess loss, as the modal field in lower-index cladding material within the notch has a larger phase velocity than that in the core. This helps bend the planar modal phase front of the stem to better match the phase front on the bent arms.

14.5   PROPAGATION ANALYSIS

In the weak-guidance approximation, propagation is determined by the scalar wave equation. This approximation applies to the guided fundamental mode of BCWs and devices, but is a valid description of radiation loss only for forward and close-to-forward directed radiation. This constraint is implicit in the formulation of the weak-guidance approximation, which assumes that the local wave vector is approximately parallel to the waveguide axis. For the small-angled taper and small bend angles in the *Y*-junction of Figure 14.6, this requirement is satisfied.

There are no purely analytical methods for accurately analysing propagation through *Y*-junctions, for, as we saw in Section 4.1, the scalar wave equation does not have an analytical solution for square- or rectangular-core waveguides. Accordingly, we have to resort to numerical or semi-numerical techniques.

14.5.1   Coupled local-mode theory

One possible method of solution would be to use coupled local-mode theory, based on the propagation of the fundamental local supermode through the *Y*-junction (Weissman *et al.*, 1989; McIntyre and Snyder, 1974). To lowest order, this mode propagates along the stem and outward along the arms without loss of power, i.e. we get the perfect lossless *Y*-junction. To first order, power is lost by coupling to the radiation field. Since the radiation is predominantly forward-directed and the arms of the *Y*-junction taper outwards, we would expect some of this radiation to couple back into the fundamental mode, in much the same way that recoupling occurs in tapered waveguides. In other words it would be necessary to solve the coupled equations, either exactly, or to higher order, in order to correctly account for this mechanism.

The disadvantage with this approach is the need to work with the continuum of radiation modes in order to correctly represent the radiation field. This requires the solution of a set of coupled discrete-continuum mode equations, which does not provide physical insight into the problem, and is very complicated.

14.5.2   Beam propagation method

An alternative, and more general, approach is to use the beam propagation method (BPM), as it obviates the need to work with the modes of the structure. This is a purely numerical method which works directly with the scalar wave equation, and, in the version adopted here, solves this equation in each cross-section of the *Y*-junction and then propagates the solution forward by a small increment to the next cross-section. This powerful technique can be regarded as a numerical simulation of actual propagation through the device.

As it is a forward-propagating technique, it cannot take account of any backward-directed radiation arising from reflections off the tapered sections of the Y-junction. However, this effect is expected to be very small because of the very slight difference in core and cladding indices. As was shown in Section 7.1.2, using Fresnel reflection, it is of the order of 50–60 dB lower than the transmitted power, and consequently will be ignored.

The results presented in the following section, using the BPM, were performed by employing a full two-dimensional algorithm for the cross-section. A rectangular area of sides $L_x \times L_y$ μm^2 was discretized into a grid of 128×128 points, or 256×64, depending on the core size along the Y-junction. The step size Δz in the direction of propagation was set to 2 μm and the wavelength to 1.3 μm. These parameter values were decided after a thorough examination of the geometries, propagation distances and refractive index values involved, and were found to be appropriate.

14.6   RESULTS

The calculations were based on the geometrical model in Figure 14.6. The stem is tapered outwards, approximately adiabatically, until its initial square cross-section becomes a rectangle with an aspect ratio of 2. The arms of the junction are formed by two circular arcs of constant curvature which are tangential to the uptaper. The effect of the notch is minimal for all practical cases, so for simplicity, was not included in the model. Except where otherwise mentioned, the results are based on core and cladding indices $n_{co} = 1.45$ and $n_{cl} = 1.447$, respectively, wavelength $\lambda = 1.3$ μm and core half-width chosen such that $V = 2.0$ for the square-core BCW, i.e. $\rho = 4.43$ μm. The field was propagated through the Y-junction until the separation of the arms reached 8ρ, regardless of the splitting angle θ.

14.6.1   Forward-directed propagation

When the Y-junction is operated in the forward direction, light in the stem is divided at the junction and an equal amount travels down each branch. The excess loss is due to light radiating from the guiding structure, because of the loss of translational invariance beyond the stem. The radiated light propagates away symmetrically from both sides of the junction and arms. However, because this radiation is mainly forward-directed, it stays relatively close to the arms of the Y-junction for some distance, especially if the branching angle θ is small. The propagation of the radiation and the guided modes is approximately

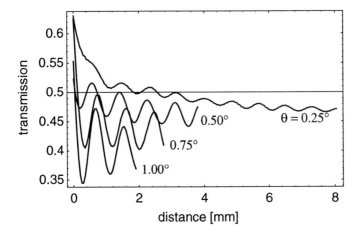

Figure 14.7 Fraction of powerin each arm of the *Y*-junction as a function of distance *z* defined in Figure 14.1.

parallel in the arms, and this causes a beating effect. After sufficient distance far along the arms, this effect gradually disappears and only the guided power remains in each arm.

The distance required for the beating to become insignificant is relatively large, and this presents a limitation on the numerical calculation. The BPM algorithm is based on a discretization of the transverse field on a finite grid. This grid cannot be made arbitrarily large as spacing of the grid points is directly related to the accuracy of the method. As the arms of the *Y*-junction diverge, they move closer to the boundary of the grid, and, if the beating has not disappeared, the final calculation of power involves both the guided and radiated power.

To illustrate this effect, consider the simple case of the linear *Y*-junction in Figure 14.1. The separation between the cores in the branches is given approximately by $L = 2z/\theta$, where θ is the half-branching angle and z the distance measured from the initial split of the arms. A BPM simulation was undertaken with the fundamental mode of the stem excited. The results are presented in Figure 14.7, where transmission is defined as the fraction of initial power in each arm as a function of distance z. It is plotted for different values of the half-branching angle θ. The angle of the arms, relative to the z-axis of the

Y-junction, is taken into account when calculating the overlap with the fundamental mode of each arm, to correctly determine the power propagating.

The strong beating between the guided and radiated power, which is plainly evident in each of the four curves for different values of branching angle, decreases slowly with z. The amplitude of the beating increases with increasing branching angle, in keeping with increased radiation with increasing branching angle, as predicted by the delineation criterion (14.8). Further, the period, or beat length, is approximately independent of the branching angle. This result is consistent with the definition of beat length between the fundamental mode in the arm and the approximately forward-directed radiation field (Love, 1989)

$$L_b = \frac{2\pi}{\beta - k n_{\mathrm{cl}}} \approx 2\pi \sqrt{\frac{2}{\Delta} \frac{V}{W^2}} \rho$$

If we set $\Delta = 0.003$, $V = 2.0$, and $\rho = 4.44$ μm, so that $W = 1.44$, then the beat period $L_b \approx 700$ μm, in agreement with the results of the simulation.

Propagating light in the forward direction of any symmetric Y-junction generally produces similar interference behaviour. For the more complicated designs, such as the ones proposed in Section 14.5, this phenomenon will still be presentand will make an accurate assessment of loss difficult. The curved-arm design has the advantage that the radiated light will propagate roughly parallel to the z-axis, whereas the fundamental-mode power in the arms will be continuously steered away from this direction. An alternative strategy for avoiding this problem altogether is now addressed.

14.6.2   Backward-directed propagation

If we reverse the direction of propagation in the Y-junction by exciting the fundamental mode in one of the arms, then, as we showed in Section 14.1, after propagating through the junction, half of its power remains in the fundamental mode of the stem and the reminder is radiated away from the stem. In contrast to the forward-propagating situation discussed above, where the radiated power remains in the proximity of the arms for a considerable distance, the power radiated from the stem diverges relatively quickly away from the stem. This means that, by using the BPM to propagate the fundamental mode on the arm through the Y-junction, a much more accurate value of guided power in the stem can be determined using only a short length.

Exciting the Y-junction in the backward direction raises the question of reciprocity. It has been proved that the total excess loss calculated for this situation is equal to the loss in the reciprocal situation of forward-directed propagation (Baets and Lagasse, 1982). In the following discussion, we use this result and assume that the Y-junction loss is also reciprocal.

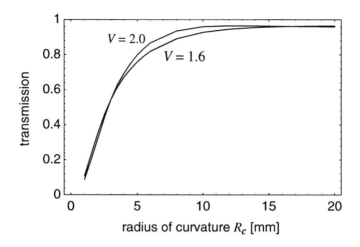

Figure 14.8 Power transmission fraction for the curved-arm *Y*-junction as a function of the radius of curvature of the arms for $V = 2.0$ and $V = 1.6$. The constant V-value is maintained by varying the core half-width ρ.

Figure 14.8 shows the fraction of power remaining in the stem of the *Y*-junction, when operated in the backward direction, as a function of the radius of curvature of the arms for two different V-values. In other words, unit transmission corresponds to 50% of the power being in the incident fundamental mode in the arm. As the radius of curvature decreases, the device becomes shorter for the given branch separation of 8ρ, but at the same time it becomes more lossy. In the limit where both arms are straight, the loss approaches 4.1% (0.2 dB) for $V = 2.0$ and 3.6% (0.18 dB) for $V = 1.6$ due to the effects of the junction and the following downtaper. The higher loss corresponds to the larger V-value, i.e. to the more tightly confined modal field. These values are in good agreement with experimental results (Takahashi *et al.*, 1991). In each case, the loss starts to increase rapidly when the radius of curvature decreases below 10 mm for $V = 2$, or 15 mm for $V = 1.6$, in keeping with the increased bend loss for the smaller V-value.

As we have already pointed out, curved arms help reduce the overall device length. In order to quantify this effect, we define the device length L_d as measured from the base of the tapering region of the stem to the end of the arms illustrated in Figure 14.6. Figure 14.9a then plots the device length L_d versus the radius of curvature R_c of the arms for the parameter values used in Figure 14.8. In the limit where $R_c \rightarrow \infty$, i.e. the straight arm limit, the overall *Y*-junction is approximately 4.0 mm long, but is reduced to around 1.3 mm

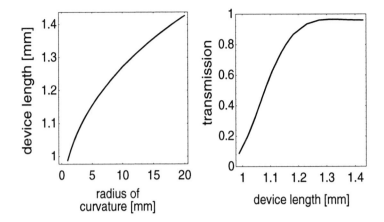

Figure 14.9 (a) Total$Y$-junction length L_d as a function of the radius of curvature. (b) Power transmission fraction as a function of Y-junction length.

when these arms have a 10 mm radius of curvature. Figure 14.9b plots the transmitted power as a function of device length for the arm parameter values, and shows that, at an arm radius curvature of 10 mm, there is virtually no additional loss due to bending.

REFERENCES

Baets, R. and Lagasse, P. E. (1982) Calculation of radiation loss in integrated-optics tapers and Y-junctions. *Applied Optics*, **21**, 1972–1978.

Ladouceur, F. (1991) *Buried channel waveguides and devices*. Ph.D. thesis, Australian National University.

Love, J. D. (1989) Application of low-loss criterion to optical waveguides and devices. *IEE Optoelectronics Part-J*, **136**, 225–228.

Love, J. D. and Henry, W. M. (1986) Quantifying loss minimisation in single-mode fiber tapers. *Electronics Letters*, **22**, 912–914.

McIntyre, P. D. and Snyder, A. W. (1974) Power transfer between nonparallel and tapered optical fibers. *Journal of the Optical Society of America*, **64**, 285–288.

Stewart, W. J. and Love, J. D. (1985) Design limitation on tapers and couplers in single-mode fibre. Pages 559–562 of: *Proceedings of the 5th European Conference of Optical Communication*.

Takahashi, H., Ohmori, Y. and Kawachi, M. (1991) Design and fabrication of silica-based integrated-optic 1×128 power splitter. *Electronics Letters*, **27**, 2131–2133.

Weissman, Z., Hardy, A. and Marom, E. (1989) Mode-dependent radiation loss in Y junctions and directional couplers. *IEEE Journal of Quantum Electronics*, **QE–25**, 1200–1208.

15
Summary

In the preceding chapters, we have provided qualitative physical insight into the propagation of light along planar waveguides, the nature of the various loss mechanisms, and the concepts behind the simplest planar optical devices. We have also presented a variety of analytical and numerical techniques for analysing propagation, concentrating mainly on single-mode buried channel waveguides, but also taking into account propagation losses due to splicing, surface roughness, substrate leakage and bending loss. We have shown how to analyse and design a variety of simple, single-mode passive devices with specified operating characteristics and low excess loss, as well as optimally low-loss curved waveguide paths connecting the output and input ports of successive devices in an optical circuit.

Putting the physical and analytical techniques together, they provide a basis for the accurate design of compact, low-loss planar devices and arrangements of devices, connecting waveguide paths and pig-tailed fibres, as well as generating an estimate of the expected overall loss. Whilst beyond the scope of the present book, the generalization of these modal techniques to multimode planar waveguides and associated devices, such as couplers and splitters, is a straightforward extension of the results presented here. Furthermore, their analysis can be augmented by standard ray-tracing techniques (Snyder and Love, 1983, Part I).

The incorporation of nonlinear and active materials within silica-based planar technology has been demonstrated by several research groups, and the integration of semiconductor sources and detectors into silica-based optical circuitry is also being addressed. Each of these developments presents new physical, analytical and design challenges, but are necessarily built onto an understanding of the basic waveguiding concepts presented in this book.

REFERENCES

Snyder, A. W. and Love, J. D. (1983) *Optical waveguide theory.* London: Chapman & Hall.

Index